The State of the Universe

The State of the Universe

Wolfson College Lectures 1979

EDITED BY
GEOFFREY BATH

CLARENDON PRESS OXFORD

1980

Oxford University Press, Walton Street, Oxford OX2 6DP

OXFORD LONDON GLASGOW
NEW YORK TORONTO MELBOURNE WELLINGTON
KUALA LUMPUR SINGAPORE JAKARTA HONG KONG TOKYO
DELHI BOMBAY CALCUTTA MADRAS KARACHI
NAIROBI DAR ES SALAAM CAPE TOWN

British Library Cataloguing in Publication Data

The state of the universe.—(Wolfson College.
  Lectures; 1979).
  1. Astronomy
  I. Bath, Geoffrey   II. Series
520   QB43.2   79-41393

ISBN 0-19-857549-1

Filmset in 'Monophoto' by
Eta Services (Typesetters) Ltd., Beccles, Suffolk

Printed and bound in Great Britain by
Morrison & Gibb Ltd, London and Edinburgh

Preface

Now my own suspicion is that the universe is not only queerer than we suppose, but queerer than we can suppose.

J. B. S. Haldane

Possible worlds and other papers Chatto & Windus, London (1927)

The State of the Universe developed from a series of lectures held at Wolfson College, Oxford in the Spring of 1979. The aim of the series was to survey some of the most exciting areas of astronomy of the past decade, every topic being presented by an expert in the field.

For their help in making the lectures possible I would like to thank the President and Fellows of Wolfson College, and also Gillian Moore, who handled the administrative and secretarial tasks so capably. My thanks also go to Mr Boddington and Dr Wickett for their support with the technical arrangements required at each lecture.

Wolfson College, Oxford   G.T.B.
June, 1979

Contents

Introduction   1
G. T. Bath, *Research Fellow, Wolfson College and Department of*
Astrophysics, University of Oxford.

1. The origin of the Universe   3
   D. .W. Sciama, *Fellow of All Souls College, and Department*
   of Astrophysics, University of Oxford and Professor of
   Physics, University of Texas, Austin.

2. Galaxies and their nuclei   16
   M. J. Rees, *Plumian Professor of Astronomy and Experimental*
   Philosophy, University of Cambridge.

3. The origin of the elements   40
   R. J. Tayler, *Professor of Astronomy, Astronomy Centre,*
   University of Sussex.

4. The stars as suns   68
   D. E. Blackwell, *Savilian Professor of Astronomy, University of*
   Oxford.

5. The X-ray Universe   93
   K. A. Pounds, *Professor of Physics, Department of Physics,*
   University of Leicester.

6. Black holes   121
   R. Penrose, *Rouse Ball Professor of Mathematics, University of*
   Oxford.

7. Planetary exploration   144
   Garry E. Hunt, *Head of Laboratory for Planetary Atmospheres,*
   University College, London.

8. New ways of seeing the Universe   181
   F. G. Smith, *Director, Royal Greenwich Observatory.*

Index   197

Introduction

G. T. BATH

Over the past two decades astronomy has undergone a series of remarkable developments. This is mainly due to the impact of modern technology on a subject which in the past has been tied to the earth by gravity, and to the visible region of the electromagnetic spectrum by the transmission properties of the Earth's atmosphere. The discovery of radio emission within our own Galaxy before the Second World War marked the beginning of exploration away from visible wavelengths, and the launch of rocket- and balloon-borne instruments, and later of scientific satellites allowed escape from the Earth's surface to an observing-point where the appearance of the sky over the whole electromagnetic spectrum was opened up—from the microwave region, through infrared, ultraviolet, and X-rays, to the gamma-ray region.

These explorations have brought to light a whole new Universe. It is as if man were suddenly able to see the Universe simultaneously both in colour and in focus—moving about freely, picking out features of interest, examining them from a variety of points of view. Previously, tied to the visible part of the spectrum, the Universe was seen through the filter of rose-coloured spectacles, full of monochromatic details but tantalizingly lacking in perspective.

As a result of these observational developments astronomy has come of age. Once this leap is over, it is hard to imagine astronomers experiencing again the same sense of revelation, of tearing aside veils which covered so many fields of research and so many objects in the sky. Whole classes of new, strange, and unpredicted objects have been discovered, ranging from the distant and energetic quasars, to the intense binary X-ray stars within our own Galaxy.

One important consequence of this burst of new knowledge is that astronomers are now on the threshold of understanding the structure and evolution of galaxies. This parallels the theory of the structure and evolution of stars that resulted from the work of Eddington, Chandrasekhar, Schwarzschild, Hoyle, and others in previous decades. With galaxies all the essential data are now becoming available, and understanding of the relationship between different classes of galaxy is just beginning to emerge.

The X-ray region has helped this programme with studies of active galaxies, quasars, and the distribution and composition of gas in clusters of galaxies. It is also becoming clear that the galactic X-ray sources fill an important gap in our knowledge of stars. Apart from being the most likely objects in which a black hole might be detected with certainty, the X-ray binaries are providing clues about the nature of important classes of optical stars whose properties were previously mysterious. In fact almost all areas of astronomy—star formation and evolution, supernova remnants, molecular clouds, galactic structure, galaxy clusters, and the origin of the Universe and the matter within it—these, and many more have seen major changes over the past two decades.

Although much of the excitement has been generated by the application of new observing techniques, this is not the whole story. The popular image of astronomers as men sitting at the eyepiece of a telescope waiting for the Universe to reveal itself is as inappropriate as the image of Sherlock Holmes waiting in his Baker Street rooms for Moriarty to come and confess. It is true that astronomers spend much of their time gathering information through telescopes, whether optical or otherwise, or exploring mathematical solutions to established problems. But the real effort is spent attempting to formulate new questions whose answers could lead to a fuller account of the present state of the Universe. This is the life-blood of astronomy, as much as the discovery of new classes of physical object in space. As Einstein and Infeld wrote, 'The formulation of a problem is often more essential than its solution.' Several of the chapters in this book are more concerned with the formulation of problems and with suggested or partial solutions, rather than with established facts and simple answers.

In addition to developments in observational astronomy, technical progress has brought about a revolution in studies of the solar system. Space missions to the inner planets, and recently to the outer planet Jupiter by *Voyager 1* and *Voyager 2*, have brought the field of planetary astronomy directly back to Earth. The traditionally Earth-oriented sciences of meteorology, seismic physics, geophysics, geology, and atmospheric physics are all being enriched now that the environments of other planets can be compared with the Earth.

The growth of astronomy over the past two decades has generated widespread popular interest. This interest reflects Man's concern with his place in the Universe. It is important to recognize that as Man makes progress towards a physical understanding of the stars and galaxies, of the space between them, and of the Universe at large, understanding is being achieved through the collective efforts of individual men and women. This book and the lectures out of which it grew, attempts to paint a broad picture of our present understanding of the state of the Universe. Each chapter outlines one man's view of recent progress in fields of astronomy that have seen major changes in recent years. The authors have taken significant steps themselves on the path of discovery. Each has his own interest and style, and I hope this book will be of value as much for that reason as for the queer discoveries it relates.

1

The origin of the Universe

D. W. SCIAMA

Introduction

Perhaps the most important scientific discovery of the twentieth century is that the Universe as a whole, considered as a single totality, is amenable to rational enquiry by the methods of physics and astronomy. These methods have revealed much about the basic large-scale structure of the Universe in both space and time. In space we can say that the ultimate building-brick of the Universe is a cluster of galaxies. The scale of this structure is remarkable—it would take light several million years to cross a typical cluster of galaxies, and these clusters are distributed throughout space to the furthest reaches of the observable Universe. Using the most powerful existing telescopes we can register these clusters out to distances exceeding five thousand million light-years.

But even more remarkable is the structure of the Universe in time. It is this structure which accounts for the dramatic impact which studies of cosmology have had on the general culture of our time. One might have thought *a priori* that while change occurs in localized regions of the Universe, such as in our own Milky Way galaxy, these changes would occur against a backdrop of an unchanging large-scale Universe. The most important single discovery about the Universe in our century is that this is not so. Because of its systematic expansion, we have to say that *the Universe as a whole is a system evolving in time*, so that we can speak of earlier or later cosmic epochs.

Moreover this change has, from an astronomical point of view, to be regarded as occurring remarkably rapidly. When life first appeared on the Earth about three thousand million years ago, the galaxies were so much closer together that the mean density of the whole Universe was about twice as great as it is today. And, if that does not seem to be dramatic, I would add that, according to the simplest extrapolation, only about three times further back into the past the material particles in all the galaxies were essentially superposed on one another, so that the density of the Universe was infinite. This corresponds to the famous 'Big Bang' origin of the Universe. The observational and theoretical evidence in favour of the Big Bang, and the question whether an actually singular origin of the Universe can be avoided, will occupy us in this chapter.

FIG. 1.1 A spiral galaxy seen edge-on in Canes Venatici. (Photograph from the Hale Observatories.)

The world of galaxies

In order to prepare for this discussion we consider first the main facts about galaxies. Pictures of a few of them are shown in Figs. 1.1, 1.2 and 2.1 to 2.5. Of particular interest is Fig. 1.2, which shows the famous galaxy in Andromeda. This galaxy, which can be seen by the naked eye, is our nearest neighbour in

FIG. 1.2 The spiral galaxy M31 in Andromeda, a member of the Local group. (Photograph from the Hale Observatories.)

space amongst the large galaxies. It was the first such object to be shown to lie outside the confines of the Milky Way—this was one of the achievements of the great American astronomer Edwin P. Hubble, and dates from 1924. In addition to its historical role this galaxy is of interest because astronomers believe that its structure is similar to that of our own Galaxy. Both are spirals of similar size and shape, although Andromeda appears to be somewhat larger than the Milky Way. It was shown by Harlow Shapley in 1918 that the Sun does not lie at the centre of our Galaxy but close to its disc a long way out (Fig. 3.2).

Galaxies come in a variety of shapes and sizes. It will suffice for our purposes if we quote the main properties of our own Galaxy, which is typical of the larger members of the population:

> *Mass* $\sim 10^{44}$ grams (corresponding to $\sim 10^{11}$ stars like the sun)
> *Radius* $\sim 10^{23}$ centimetres ($\sim 10^5$ light-years)
> *Rotation period* $\sim 3 \times 10^8$ years
> *Age* $\sim 10^{10}$ years

These are properties of individual galaxies. We shall also need a number which specifies the spacing of galaxies. As a very round average we may take

$$\text{Separation} \sim 10^{25} \text{ centimetres } (\sim 10^7 \text{ light-years})$$

so that in this average sense galaxies are separated by about one hundred times their own size. Relative to their size galaxies are closer together than the stars of our own Galaxy, which are separated on average by more than ten million times their own size.

Our figure for the separation of galaxies is very much an average, because galaxies are far from being uniformly, or even randomly, distributed in space. Many of them occur in clusters ranging in size from pairs, through small groups, such as the Local Group which contains the Milky Way and is believed to consist of about twenty galaxies, out to the giant clusters, some of which contain over a thousand galaxies (Fig. 1.3). There is some evidence that the clusters are themselves super-clustered, but this is a refinement which we need not be concerned with here. Later on we shall describe the evidence that on a very much larger scale the Universe is remarkably uniform.

The recession of the galaxies

The huge scale of the Universe in space is difficult to comprehend—indeed, as we shall see, it may be infinitely large. Nevertheless the conceptual challenge of modern cosmology stems mainly from the still more remarkable feature that the Universe is in a *dynamic* state. Clusters of galaxies are not on average at rest with respect to each other, but are moving apart. One can tell this by observing the optical spectrum of a galaxy. This spectrum is crossed with dark lines arising from absorption processes by atoms of

FIG. 1.3 The Coma cluster of galaxies. (Photograph from the Hale Observatories.)

various chemical elements (or, in exceptional cases, by emission lines). The elements giving rise to these lines can be identified from the patterns of wavelength involved. When one does this one finds that for most galaxies the lines have somewhat longer wavelengths than for the same lines produced in the laboratory. This red-shift is attributed to the Doppler effect, that is, to a recession motion of the galaxy with respect to our own. For small velocities the Doppler shift is given by

$$\frac{\lambda_{\text{obs}} - \lambda_{\text{rest}}}{\lambda_{\text{rest}}} = \frac{v}{c},$$

where λ_{obs} and λ_{rest} are the observed and rest wavelengths respectively, v is the velocity of the galaxy (taken positive for a motion of recession) and c is the velocity of light. For velocities close to that of light this relation must be modified to allow for relativistic effects.

In 1929 Hubble announced that the red-shifts of the galaxies followed a simple pattern which we now call Hubble's law. According to this law the velocity of recession of a galaxy is simply proportional to its distance, r. We can write this law

$$v = Hr,$$

where the constant of proportionality H (Hubble's constant) is currently believed to be given, within a factor of two, by

$$H = 100 \text{ kilometres per second per megaparsec.}$$

In other words, for every megaparsec of its distance (a megaparsec is about three million light years) a galaxy has a recession velocity of 100 kilometres per second.

For some purposes it is more convenient to write the constant of proportionality of the Hubble law in the denominator, since it then has the dimensions of time. We can thus write

$$v = \frac{r}{\tau}$$

where τ (also often called the Hubble constant) is given, again to within a factor 2, by

$$\tau = 10^{10} \text{ years.}$$

Hubble's law was first established for relatively nearby galaxies, but it has stood the test of time remarkably well. Fig. 1.4 from Allan Sandage's 1968 Halley lecture in Oxford shows a more recent compilation of data. The

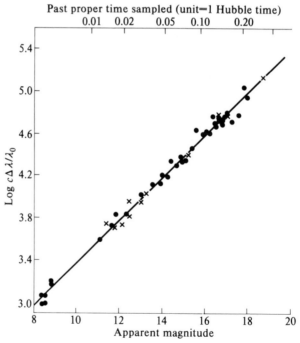

FIG. 1.4 The Hubble law, showing the linear relation between recession velocity and distance. (Courtesy of *Observatory*.)

relation between velocity and distance remains linear out to much greater distances (Hubble's original data are all at values less than 3.0 in the recession velocity.

Immediately after Hubble's law was announced it was appreciated that it had two profound implications. The first is that it is the only law which has the property that an observer on any other galaxy looking out at the Universe would see the *same* law for its expansion. In other words, we on the Milky Way are not to be regarded as at the unique centre of the expanding system of galaxies. Moreover we are on a *typical* galaxy of a population, all of whose members play equivalent roles in their relation to the structure of the Universe. We may thus say that just as Copernicus dethroned the Earth and Shapley the Sun, so Hubble dethroned the Milky Way.

This conclusion, while striking, could not be said to be against the spirit of the time. What was more striking, and to many unpalatable, was the second conclusion which was drawn from the form of Hubble's law. If the galaxies are moving apart today, then in the past they were closer together, and one can estimate how long ago all the galaxies (or, better, the pregalactic material particles) were on top of one another. An accurate calculation would require one to know whether the motion of each galaxy is appreciably decelerating or accelerating, whereas in practice we can observe each galaxy only at one instant. However, for a rough estimate we may ignore this effect and assume that each galaxy has always had the velocity we now observe it to have. In the absence of any unsuspected effect we would thereby expect to make an error of less than a factor 2. Thus came the astonishing result which so disturbed Hubble's contemporaries: the density of the Universe was infinite just a time τ in the past, that is, a mere 10^{10} years ago. This result seemed even more dramatic then than it does today, because at the time it was thought that τ was only 2×10^9 years, so that the Universe actually appeared to be 'younger' than some of its material contents such as the Earth (age $\sim 4 \times 10^9$ years) and the Sun (age $\sim 6 \times 10^9$ years). More recent estimates of τ have resolved this time-scale difficulty, but it remains true that the Universe had a totally different structure, essentially an origin in a Big Bang, at a time in the past only about twice as great as the formation period of the Earth and the Sun.

Various attempts have been made to evade this conclusion, of which the most attractive is the steady-state hypothesis proposed by Hermann Bondi, Thomas Gold, and Fred Hoyle in 1948. According to this hypothesis, matter is being continually created in the Universe at such a rate that its mean density remains constant despite the expansion. The large-scale properties of the Universe would then be the same in the distant past and in the distant future as they are today, and there would be no question of a beginning (or an end) to the Universe. We shall see shortly that this hypothesis has been disproved by recent observational data.

Before discussing these data we should say something about the likely future of the Universe. Discussions of this question must depend heavily on theory, and we shall use the best available theory, namely Einstein's general relativity (although one can go quite far also using Newtonian models of the Universe). One would expect the rate of expansion of the Universe to be slowing down as a result of the gravitational action of the galaxies on each other, so everything depends on the importance of this effect in relation to the present rate of expansion. It is a good approximation to smooth out the contents of the Universe to a uniform distribution and to work in terms of the resulting average density. There exists a critical value for this density such that if the actual average density of the Universe is less than critical then the Universe is destined to expand for ever. In such a case the volume of the Universe at any one instant would be infinite. On the other hand if the actual density is greater than critical then the gravity of the Universe is so strong that the expansion would be brought to a halt, to be followed by a collapse into what is sometimes called the 'big crunch'. In this case the volume of the Universe at any one instant would be finite. At the critical density itself the Universe would expand for ever (but only just), in the sense that the rate of expansion would tend to zero in the infinite future. For this intermediate case the volume of the Universe would be infinite.

According to general relativity the critical density ρ_{crit} is given by

$$\rho_{crit} \sim 10^{-29} \text{ grams per cubic centimetre}$$

for a Hubble constant $\sim 10^{10}$ years.

If one smooths out the known material in the galaxies one obtains an average density ρ_{gal} which is believed to be given by:

$$\rho_{gal} \sim 10^{-30} \text{ grams per cubic centimetre}.$$

Uncertain as this quantity is, most astronomers do not think that it could be increased by as much as a factor of 10. So, if general relativity is correct, our conclusion must be that the Universe is infinite and will expand for ever, with energy to spare, unless astronomers can discover a form of energy density between the galaxies which is ten times greater than the density arising from within the galaxies.

This 'missing matter' problem has attracted a good deal of attention. There are various forms the missing matter could take without having yet been observed, such as bricks, black holes, neutrinos, or gravitational waves. With the possible exception of black holes, none of these possibilities seems very plausible. The most attractive possibility would be a hot intergalactic gas, but recent X-ray observations make it unlikely that such a gas could have the critical density. The final state of the Universe thus remains for the time being unknown.

The cosmic black body background

The state of affairs described so far (with the exception of the value of the

Hubble constant) was well established by the mid-1930s. There followed a rather quiescent period for cosmology, both observationally and theoretically, which was punctured in 1948 by the introduction of the steady-state theory and in the 1960s by various observational developments. Of these by far the most important was the accidental discovery in 1965 of evidence for the existence in the Universe of a black body radiation field at a temperature of about three degrees above absolute zero. The discoverers of this radiation field, the radio astronomers Arno Penzias and Robert Wilson, were awarded the Nobel prize for this discovery just recently.

The observations to date, which are shown in Fig. 1.5, indicate that this radiation field has the (Planck) spectrum we would expect if it had come into thermal equilibrium with matter at a definite temperature. Further observations show that the intensity of the radiation is the same in all directions to a precision exceeding one part in a thousand. This strongly suggests that the radiation is associated with the Universe as a whole, and represents by far the most accurate measurement ever made in cosmology. In particular it tells us that the Universe is very uniform on a large scale, since non-uniformities would disturb the isotropy of the 3 K background.

To understand the further significance of these observations one must appreciate that a thermal radiation field filling the Universe would cool down as the Universe expands. The relation involved is a simple one: when the distance to another galaxy doubles the temperature of the radiation is

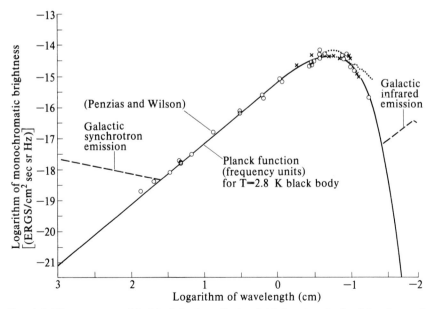

FIG. 1.5 The spectrum of the black-body radiation field showing the fit of the observed points to a Planck spectrum at 2.8 K. (Courtesy of the American Philosophical Society.)

halved. Similarly, if one goes into the past, when the distance to another galaxy was half the present distance, the temperature of the radiation was double. Thus, unless the radiation was introduced at a relatively late stage, we can say that the early dense Universe was hot, and we have the concept of the 'Hot Big Bang' origin of the Universe, introduced in 1948 by George Gamow, Ralph Alpher, and Robert Herman, and revived independently in 1965 by Robert Dicke, Jim Peebles, Peter Roll, and David Wilkinson.

Mention of 1948 reminds one that that was also the year of the steady-state theory. Why are we here taking for granted that there was a Big Bang origin of the Universe? The reason is that the existence of the 3 K radiation field itself represents powerful evidence that the Universe was once much denser than it is today. This is incompatible with the steady-state theory. This follows from the fact that in a Universe as dilute as the present one, there would not be enough time for a radiation field to come into thermal equilibrium with matter. Thus in the steady-state theory no explanation is provided for the presence of an equilibrium radiation field. By contrast, in evolving models of the Universe the thermalization time is shorter than the time-scale of expansion in the early dense stages. According to the standard theory the last moment at which thermalization was possible in the time available occurred about 300 years after the Big Bang, when the density of the Universe was about 10^{15} times greater than it is today and the temperature was about a million degrees. Thus in observing the 3 K background radio astronomers are observing the result of processes which occurred 300 years after the Big Bang at the latest. There is a variant of the standard theory in which thermalization was produced later by molecules and dust generated by a pre-galactic population of stars, but even this process would have occurred about 3 million years after the Big Bang, much earlier than the present age of 10 thousand million years.

The 3 K background thus provides us with a means of studying the behaviour of the Universe in its hot early epochs. Unless the radiation was itself produced at a relatively late stage, the Universe would have been hotter still at epochs before the last moment of thermalization, and the question arises whether there exist today fossil records of processes which occurred at these higher temperatures. This appears to be the case. We find in the Galaxy today that about one atom in ten is helium, the rest being mainly hydrogen, with a small admixture of heavier elements. It is believed that these heavier elements are cooked by nuclear reactions promoted by the high temperatures in the interiors of stars and the heaviest elements in supernova explosions, but it does not seem possible to produce more than about ten per cent of the observed helium in this way. By contrast, the conditions prevailing in the early Universe are suitable for producing abundant helium but very little of the heavier elements. The key moment occurred when the temperature was a thousand million degrees, the density 10^{27} times the present density and the time a mere one hundred seconds after the Big Bang. Calculations by Jim Peebles and by Bob Wagoner,

William Fowler, and Fred Hoyle show that the amount of helium formed at that time corresponds closely to the abundance which we observe today. If this explanation is correct, we can say that by observing helium today we are observing the result of processes which occurred only one hundred seconds after the Big Bang.

In the last few years it has become possible to construct a similar argument for the origin of deuterium, the heavy isotope of hydrogen. Recent observations have shown that there is about one atom of deuterium in the Galaxy for every 10^5 atoms of hydrogen. It is difficult to understand how this deuterium could be manufactured in stars; indeed in stellar interiors deuterium would be destroyed by nuclear reactions rather than produced. Again cosmologists have turned to the hot early Universe for a solution to this problem. When hydrogen is converted into helium at that time, a small amount of residual deuterium survives the nuclear reactions. The amount surviving is the greater the *lower* the density of matter, since with more matter there would be more nuclear reactions and the buildup to helium would be more complete. If we demand that the deuterium–hydrogen ratio produced should agree with the present observations, we arrive at a definite value for the density of the Universe. When this density is carried forward to the present, we obtain

$$\rho \sim 2 \times 10^{-30} \text{ grams per cubic centimetre.}$$

This agrees remarkably well with the value of ρ_{gal}—the density arising just from galaxies. This argument would imply that there is essentially no missing matter, and that the Universe is infinite and destined to expand for ever. There are still some loopholes to be closed before this argument can be accepted. For example, deuterium-forming processes in the Galaxy may be discovered. Less deuterium would then have been formed in the early stages, leading to a denser Universe. Alternatively, the 'missing matter' could be in a form in which it did not partake of the early nuclear reactions, for example, if it were already in black holes. Such matter would not lead to the destruction of deuterium, but might still be sufficient to lead to a 'big crunch'. The future of the Universe thus remains uncertain.

It clearly makes sense to discuss the processes which occurred only one hundred seconds after the Big Bang. The nuclear reactions which built up helium and deuterium are well understood. Indeed at the time of one hundred seconds all the prevailing physics can be regarded as known.

Does it also make sense to go back further still? At much earlier times we enter a regime where the physics is not well understood. Nevertheless we cannot simply exclude such a regime from present consideration, if only because processes which occurred then may have left their mark on the Universe as it is today. For example, the origin of the heat now in the 3 K background may be associated with quantum processes which occurred at a time of 10^{-23} seconds, when the Hubble velocity across a distance equal to the radius of an elementary particle was close to the velocity of light, or even at a time of 10^{-43} seconds, when we believe that a (still undeveloped) quantum theory of gravity is needed to describe what was going on. Research into these problems is active at the present time, but the ultimate outcome of this work is quite unknown.

People are also beginning to grapple with the deeper problem. What happened at zero seconds, in other words at the Big Bang itself? According to the simplest models of the Universe provided by general relativity the density of the Universe was infinite at the Big Bang. For a long time this result was thought to be a consequence of the artificially high degree of symmetry assumed in these simple models. In particular if all the particles in the models moved precisely radially with respect to any one, it is not surprising that they all coincided at a finite time in the past. Introduce some irregular motions, it used to be thought, and the singular moment of infinite density would be removed. We now know that this is not correct. As first shown by Stephen Hawking in 1966, following on some pioneering work of Roger Penrose in 1965 on the singular collapse of a massive star, self-gravitation is so strong in general relativity that the singularity at the Big Bang would not be removed by realistic irregular motions.

We face here a major intellectual crisis. General relativity, which is by far our best theory of space, time, and gravitation, breaks down at the beginning of the Universe (and also in the final stages of collapse of a massive star). This breakdown prevents one from considering questions such as, what happened before the Big Bang? Was there a previous collapse of the Universe which was followed by a bounce to the present expanding phase?

These questions cannot be answered within general relativity as we know it today. To make progress we need to eliminate the singularity. According to one school of thought the way to do this is to bring in quantum effects, which are not referred to in the Penrose–Hawking singularity theorems. As we have seen, these effects are likely to be important for different reasons during the first 10^{-43} seconds and the first 10^{-23} seconds. These quantum effects are not yet understood, but there does appear to be a possibility that they might suffice to eliminate the singularity. In particular quantum effects can lead to the energy density becoming negative, in which case the associated gravity would become repulsive, thereby tending to counteract the classically attractive gravity which leads to the singularity. Whether this effect is numerically large enough to remove the singularity altogether is not known. If it is not large enough, radically new physics will have to be invented. We are here at the threshold of new ideas of fundamental importance for our understanding of the origin of the Universe.

Bibliography

Introductory books on cosmology

GINGERICH, O. (ed.) (1978). *Cosmology plus one. Scientific American Reprint*, Freeman, San Francisco.

JOHN, L. (ed.) (1973). *Cosmology now*, BBC Publications, London.

NARLIKAR, J. V. (1976). *The structure of the Universe*. Oxford University Press.

ROWAN-ROBINSON, M. (1977). *Cosmology*. Oxford University Press.

SCIAMA, D. W. (1975). *Modern Cosmology*, Cambridge University Press.

WEINBERG, S. (1977). *The first three minutes*. André Deutsch and Fontana, London.

Elementary textbooks on cosmology

BERRY, M. (1976). *Principles of cosmology and gravitation*. Cambridge University Press.

LANDSBERG, P. T. and EVANS, D. A. (1977). *Mathematical cosmology*. Oxford University Press.

More advanced textbooks include

PEEBLES, P. J. E. (1971). *Physical cosmology*. Princeton University Press.

SCHATZMAN, E. (ed.) (1973). *Cargese lectures in physics*, Vol. 6, Gordon and Breach, London.

WEINBERG, S. (1972). *Gravitation and cosmology*, John Wiley, London.

2

Galaxies and their nuclei

M. J. REES

Over four hundred years have passed since Copernicus argued that the Earth must be dethroned from the privileged central position accorded to it by Ptolemy's cosmology, and described the general layout of the solar system in the form accepted today. But the gradual abandonment of the heliocentric picture, and the full realization of the scale of the cosmos, came about much more gradually and was completed much more recently. In the eighteenth century Thomas Wright of Durham and William Herschel suggested that the Milky Way might in fact be a flat disc of stars in which the Sun was embedded. But it was not until the 1920s that—through the work of Harlow Shapley, Jan Oort, and others—the Sun lost its central position within this stellar disc; and only then was it fully realized that our Milky Way, our Galaxy, was just one fairly typical galaxy amongst hundreds of millions of others which can be registered by a large telescope. In this chapter I shall attempt to outline what galaxies are, allude to some physical processes that might determine their size and shape and control their evolution, and place them in their cosmological context. I shall briefly mention how galaxies might have formed, and how the study of galaxies may aid us in answering such basic cosmological questions as whether the Universe is destined to expand for ever or to recollapse.

Normal galaxies

The most familiar types of galaxy are the disc-like or spiral systems, such as those shown in Fig. 2.1. The stars are concentrated in a circular disc. Such galaxies seem basically to be an assemblage of stars and gas. A typical galaxy contains about 10^{11} stars, and about 10 per cent as much material in the form of gas, spread through a region about 50 000 light-years in radius. Such a system is in dynamical equilibrium, with centrifugal force balancing gravity. A typical star would orbit around the galaxy in a time of about 200 million years, its speed being 200 km/s. In the age of the galaxy there would have been time for a typical star to have completed 50 orbits. The Andromeda galaxy (Fig. 1.2), our nearest major neighbour in space, is about 2 million light-years away. Like other spiral galaxies it contains stars of all ages, and gas. Our own Milky Way would appear like this if viewed by a distant observer in the plane of the disc. The Sun lies about two-thirds of the way out towards the edge of the visible disc (Fig. 3.2). The Andromeda galaxy is similar in size to our own, but the spiral galaxies in general have a

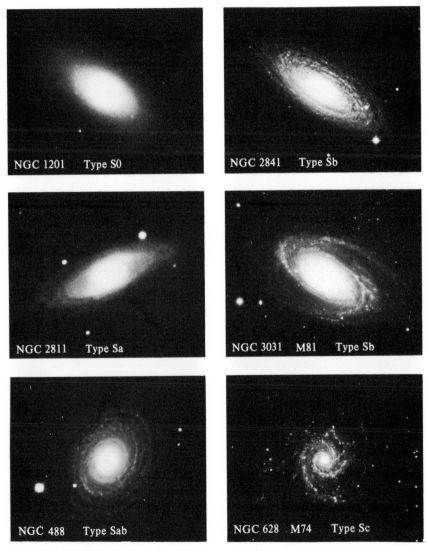

FIG. 2.1 Examples of spiral galaxies showing variation in the bulge component relative to the disc. (Photograph from the Hale Observatories.)

spread of at least 10 in mass. Fig. 2.2 shows another galaxy viewed nearly edge-on.

Even if one does not understand the physics of galaxies at all, one can classify them according to their appearance. In any such classification scheme the other main distinct class (apart from the disc galaxies) must be the 'ellipticals' (examples shown in Figs. 2.3 and 2.4). These are amorphous swarms containing between 10^{10} and 10^{12} stars, each star tracing out a complicated path under the influence of the overall gravitational field of all the other stars. The smooth brightness profile of elliptical galaxies indicates that the orbits have become thoroughly mixed in phase. An equilibrium has been established where there is a balance between the tendency of gravitation to pull the system together, and the tendency of random motions to cause the galaxy to fly apart. Even though a typical photograph does not resolve the individual stars, the stars are in fact so widely spaced relative to their actual physical sizes that there is no significant chance of collisions, or even of close encounters, over the entire 10^{10} year lifetime of such a system. Elliptical galaxies do not contain substantial amounts of gas; neither the hot glowing gas associated with regions of star formation nor the cool gas that radioastronomers detect via the 21 cm line of neutral hydrogen. Nor are any bright patches containing young blue stars observed. Elliptical galaxies,

FIG. 2.2 Spiral galaxy in Coma Berenices, viewed in the plane of the disc. (Photograph from the Hale Observatories.)

generally less photogenic than spirals, have a variety of apparent shapes. This is due not only to different orientations relative to our line of sight, but also to genuine differences in shape.

Over the last decade astronomers have developed an elaborate taxonomy for classifying galaxies with greater precision. Most galaxies can however be crudely categorized as either 'disc' or 'elliptical' systems; a simple classification can be based simply on the relative prominence of disc and elliptical components. The 'Sombrero Hat' galaxy (Fig. 2.5) displays a disc and a bulge component; most other galaxies are basically similar except that the bulge is either less or more prominent relative to the disc.

Disc galaxies contain gas *and* young stars, but elliptical galaxies contain *neither* gas nor young stars. One can try to understand this correlation between gas and young stars, and enquire whether it suggests any clues to the nature of these systems.

Let us for a moment return to our own Galaxy, where one can distinguish and directly investigate stars and gas clouds. Even though the lifetime of a star is enormously long compared to a human lifespan—so that we see, in effect, only an instantaneous 'snapshot view' of the galaxy—we can nevertheless infer a great deal about the evolution and life cycle of stars (just as someone who had never seen a tree before could infer the life cycle of

FIG. 2.3 Giant elliptical galaxy in Virgo, M87. (Photograph from the Hale Observatories.)

trees by a one-day inspection of a forest). Astrophysicists tell us that stars
are fuelled by nuclear fusion reactions in their interiors. Stars spend most of
their time on the so-called 'main sequence', where they derive their energy
from the conversion of hydrogen into helium. A star like the Sun can remain
in this phase for about 10 billion years. On the other hand, more massive
stars burn up their fuel at a much faster rate. They consequently have

FIG. 2.4 Spiral and elliptical galaxies in the same field in Virgo. (Photograph from
the Hale Observatories.)

shorter lifetimes, and eventually explode in violent supernova explosions (Fig. 3.6). In places such as the Orion gas cloud, shown in Plate 1, we can see the actual process whereby gas is being converted into stars; and we can also see large numbers of blue stars which must have formed very recently in the history of the Galaxy. When these stars die, they will eject much of their material back into the interstellar gas from which new generations can subsequently condense. The basic process going on in our Galaxy is, in effect, a cycle where gas condenses into stars. A part is subsequently returned to the interstellar gas in stellar explosions. (This is schematically indicated in Fig. 2.6). A fraction of the gas that turns into stars is, however, permanently trapped, in the sense that it gets incorporated in low-mass long-lived stars or in the compact remnants left behind after stars become supernovae. It is through this kind of recycling process that the heavy elements such as carbon, nitrogen, and oxygen in the Galaxy have been synthesized. All these elements present in the solar system must have been produced in a generation of stars which formed, evolved, and exploded before the Sun formed. The solar system subsequently condensed from gas contaminated by the debris from these earlier stellar explosions. It should not therefore surprise us that the kind of galaxies that lack gas also lack young stars.

One can now conjecture why those galaxies with no young stars and gas should be ellipticals, whereas those where star formation is still going on

FIG. 2.5 Spiral galaxy in Virgo. (Photograph from the Hale Observatories.)

should be disc-shaped. Let us suppose that all galaxies started their lives as turbulent gas clouds contracting under their own gravitation. The collapse of such a gas cloud is highly dissipative, in the sense that any two globules of gas that collide will radiate their relative energy by producing shock waves, and will merge. The end-result of the collapse of such a gas cloud, particularly if it is rotating, will be the production of a rotating disc. This is the lowest energy state that such a cloud can reach if it does not lose or redistribute its angular momentum. On the other hand, stars do not collide with each other, and are unable to dissipate energy in the same fashion as gas clouds (see Fig. 2.7). This suggests that the *rate of conversion of gas into stars* is the crucial feature determining the type of galaxy that results: elliptical galaxies will be those in which the conversion is fast, so that most of the stars have already formed before the gas has had time to settle down in a disc; contrariwise, the disc galaxies will be those in which the star-formation is delayed until the gas has already settled into a disc. According to this picture, ellipticals and spirals may have the same age. The disc galaxies are those with slower metabolisms, which have not yet got so close to the final state in which essentially all the gas is tied up in low-mass stars or dead remnants. The 'irregular galaxies' are even more extreme cases of arrested development. These are systems in which maybe less than half of the gas has so far been incorporated into stars.

The spiral arms, which are such a conspicuous feature of some disc galaxies, in fact delineate regions where star-formation is proceeding unusually rapidly. The spiral arms seem to correspond to some kind of persistent wave pattern in the disc, though there is still no completely

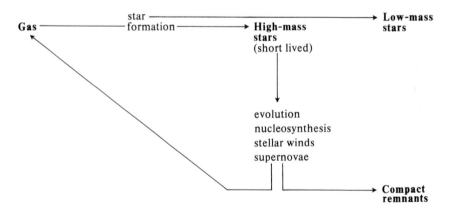

FIG. 2.6 Flow diagram showing how material is processed through high-mass stars, but trapped in low-mass stars.

satisfactory explanation of what excites and maintains the wave. The classification scheme I have outlined is somewhat confused by the existence of a class of disc galaxies (known as SO galaxies) that do not appear to contain gas. 'SO's are found preferentially in clusters of galaxies; they may be disc systems where the gas has been swept out by interactions with an external cluster medium.

Even though galaxies cannot yet be said to be well understood, there is a

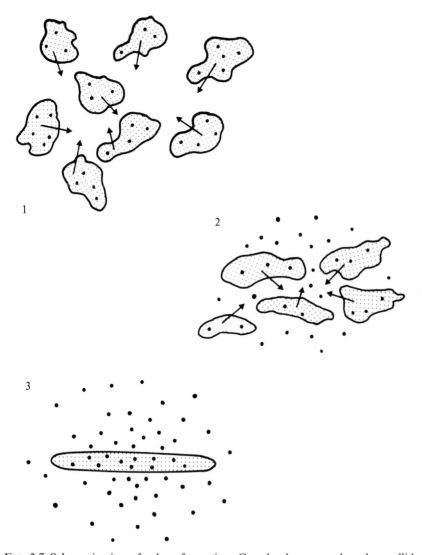

FIG. 2.7 Schematic view of galaxy formation. Gas clouds merge when they collide, whereas stars do not collide at all. The gas clouds form a rotating disc following collapse, whereas stars will be distributed throughout the volume in which they formed.

general consensus that the basic physics involved in normal galaxies is nothing more exotic or highbrow than Newtonian gravity and gas dynamics. If the cosmologist can give us primordial gas clouds of the right size and we can understand more about star formation, there seems no reason why we should not be able to explain the gross properties of present-day galaxies.

Active galactic nuclei

There are peculiar galaxies which clearly involve something more violent than ordinary stars, and where the disturbed appearance is obviously indicative of activity in the nucleus. The nearby galaxies M82 and M87 are examples of this phenomenon (Figs. 2.8 and 2.9). Even more remarkable are the so-called radio galaxies, whose power output in radio waves exceeds the total galactic luminosity of all the stars. In fact, it was the radioastronomers who provided the first clues that some galaxies might be more than just self-gravitating aggregates of ordinary stars. In 1954 Baade and Minkowski showed that the radio source Cygnus A, the second most intense object in the radio sky, was associated with a remote galaxy with a red-shift of 0.05 (Fig. 2.10). This immediately indicated that some peculiar galaxies might be detectable by radio techniques even if they were so far away that the integrated light from 10^{11} stars failed to register optically. Radio maps made with modern aperture-synthesis telescopes tell us that the emission from a

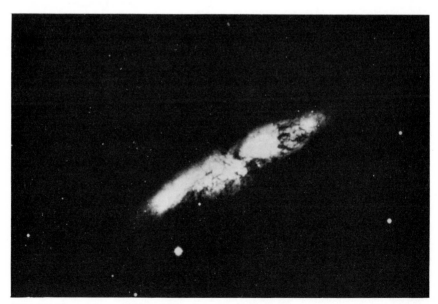

FIG. 2.8 The galaxy M82, which shows evidence of a violent explosion in the nucleus. (Photograph from the Hale Observatories.)

source like Cygnus A comes from two blobs symmetrically disposed on either side of the central galaxy. This double structure, in which the overall separation of the components may be a million light-years, or even more, seems characteristic of the strongest radio sources, and I shall return to its interpretation later. It was recognized in the 1950s that the radio emission resulted from synchrotron radiation by relativistic electrons in magnetic fields. In a celebrated paper published in 1959, Burbidge estimated that the

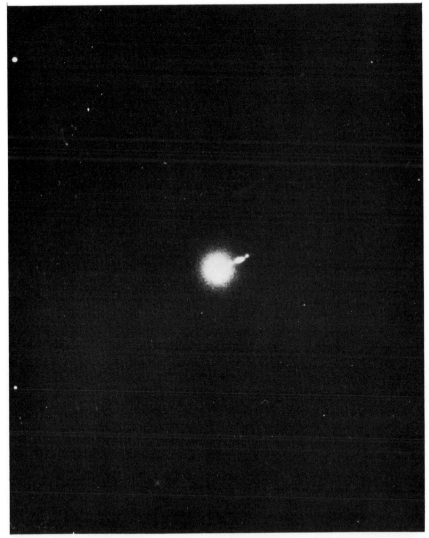

FIG. 2.9 Short exposure photograph of the elliptical galaxy M87, showing a nuclear jet. Compare with the deeper photograph of M87 in Fig. 2.3. (Photograph from the Hale Observatories.)

minimum energy content of the radio lobes of an extended source must correspond to that released by the complete annihilation of something like a million solar masses of material. This was the first indication that events occur in galactic nuclei that release energy on scales vastly exceeding even a supernova explosion, and that somehow this energy is primarily channelled into the form of relativistic plasma and magnetic fields.

The major contribution of optical astronomy to this story came in 1963, when attempts to discover the optical counterparts of some radio sources

FIG. 2.10 The optical galaxy associated with the second most intense radio source in the sky, Cygnus A. (Photograph from the Hale Observatories.)

led to the recognition that the Universe contained an unsuspected new class of objects, which looked like ordinary stars on photographic plates, but whose spectra displayed emission lines with large red-shifts. These objects, the quasars, have optical luminosities exceeding those of normal galaxies even though they are much more compact. Moreover the light seems to be emitted by the same process, synchrotron radiation, as the radio output from strong sources such as Cygnus A.

The most extreme instances of 'violent activity' are these quasars, in which a small central nucleus is outshining all the rest of the surrounding galaxy. One such quasar (known as AO 0235 + 164) was observed to undergo an outburst in the autumn of 1975 in which its rise in luminosity within a single week was equivalent to turning on 10 000 galaxies like our own. Such an object must contain a power source billions of times greater than an ordinary star, yet concentrated within a region smaller than the solar system. A consensus is gradually emerging that quasars, which obviously involve a large mass in a small space, result from some kind of runaway process whereby gas and stars accumulate in the gravitational potential well at the centre of a galaxy, until some threshold is crossed when gravity overwhelms all other effects and collapse to a massive black hole ensues. Some astronomers, including myself, are now for the first time optimistic that quasars can be understood in terms of a theory where the prime mover involves a black hole as massive as a 100 million suns fuelled by capturing gas, or even entire stars. This captured debris swirls downwards into the potential well, moving at close to the speed of light before it is swallowed. The calculations completed so far suggest that this interpretation accounts not only for the quantity of the power output of quasars but even for the form in which the power emerges.

This model may—in particular—account for the 'directivity' and double structure of radio sources. If the central engine (a black hole, surrounded by plasma moving at close to the speed of light) were surrounded by a gas-cloud, one might imagine that the relativistic plasma would tend to squirt out along the directions of least resistance. If the surrounding material were in a rotationally flattened distribution, these directions would lie along the rotation axis. One can then calculate that the only possible stationary outflow pattern is one in which the relativistic plasma makes a channel for itself whose shape is similar to the well-known 'de Laval nozzle' in jet engines. The plasma will then emerge from the galactic nucleus as two collimated beams, as shown in Fig. 2.11. The lobes of double sources like Cygnus A or 3C236 could be energized by continuously outflowing beams of plasma collimated in the nucleus; there is now firm direct evidence that these beams actually do exist.

The double radio source 3C236 is the largest such object yet discovered (Fig. 2.12). Its total linear extent is about twenty million light years. There is clear evidence of radio emission from ridges linking the components to the central galaxy. More significantly, there is now evidence that the compact source around the nucleus of the central galaxy is itself extended precisely along the direction of the overall axis of the source. This indicates not only that continuing activity has lasted for tens of millions of years, but also that the collimation direction has remained constant during that time.

Of even greater interest is the structure of the source associated with the radio galaxy NGC 6251. A low-frequency radio survey recently revealed an extremely extended double source, with very high energy content even though the current radio power output is low. A five-gigahertz map made with higher angular resolution revealed a straight jet, shown in Fig. 2.13 emerging from the galaxy and pointing towards one of the components. Very-long-baseline interferometry measurements have subsequently shown that, right in the nucleus itself, there is a source only a few light years long pointing along the jet. The nucleus of this galaxy contains a 'cosmic blow-torch', generating a jet detectable out to a distance of half a million light years, and presumably pumping energy continuously into the diffuse extended structure. The fact that the jet is seen only on one side of the galaxy might indicate that it is emerging relativistically; unless the axis is precisely in the plane of the sky, the Doppler effect would strongly enhance the detectability of the jet on the approaching side.

I will not trespass further on what is the territory of Professor Penrose's contribution. He describes the extraordinary properties of black holes. Galactic nuclei have special significance for relativists if they are indeed

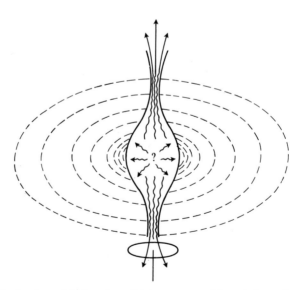

FIG. 2.11 Production of 'twin exhaust' jets as gas squirts out along the rotation axis of a rotationally-flattened gas-cloud.

places where the space of our universe gets punctured by the accumulation of large masses, which collapse, cutting themselves off from the external world but leaving a gravitational imprint frozen in the space they have left. Material that falls into such a hole can give rise to the most powerful and spectacular outbursts observed in the Universe.

Galactic nuclei and quasars are a subject in themselves. From the point of view of a galaxy, however, the quasar phenomenon can be no more than a short-lived pathological phase in the lifetime of its nucleus. If the lifetime of quasars is indeed short, dead quasars should outnumber living ones, and many earlier generations may now be defunct. A dead quasar would presumably be a massive black hole, now almost quiescent because it is starved of fuel—starved either because it is in a galaxy that is now swept clean of gas, or because it has already gobbled up all the stars near it. It is therefore interesting to ask whether massive black holes, remnants of dead quasars, may lurk in the nuclei of any well-known nearby galaxies.

The southern hemisphere galaxy Centaurus A, Fig. 2.14, at a distance of about 10 million light-years, is the nearest radio galaxy. It has very large

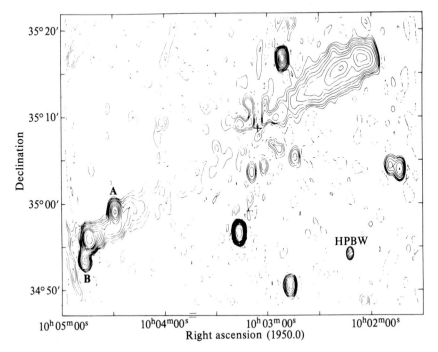

FIG. 2.12 The double radio source 3C236 with radio lobes separated by twenty million light years.

faint radio lobes which have low power output but which contain relativistic plasma with colossal total energy. It might in its prime have been a powerful radio source rivalling the quasars. One might thus expect a massive black hole to remain as the defunct remnant of the structure which gave rise to this energy. There is now known to be, right at the centre of Centaurus A, a

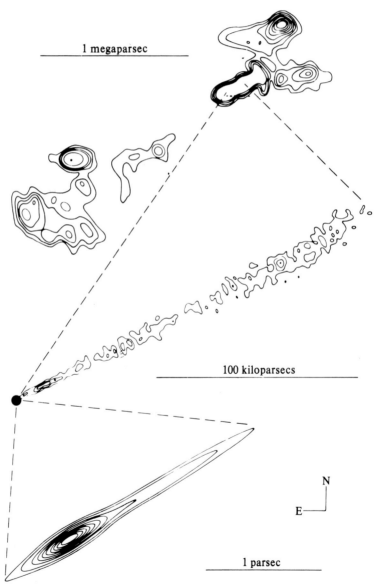

FIG. 2.13 Structure of the radio source around the galaxy NGC 6251.

tiny radio source only a light-day across. There is also an X-ray source that is variable on time-scales perhaps as short as hours. These phenomena may be attributed to a slow draining of remaining material on to a black hole, whose mass must exceed 10 million solar masses if it is the progenitor of the radio source. Centaurus A may thus contain the nearest massive black hole that is still manifesting the affects of accretion.

Slightly farther away, the giant elliptical galaxy M87 (Fig. 2.9) in the Virgo cluster has been known for sixty years to have a peculiar jet emanating from its nucleus. It is now known that the concentration of stars, and their velocity dispersion, is enhanced within the central few hundred light years. This indicates the presence of an excessive unseen mass of about five thousand million solar masses. There are several forms which this mass might take, but one obvious possibility is that it might be a single monster black hole. A black hole as large as this has the interesting property that a solar-type star could pass irreversibly within it without being tidally disrupted. There would thus not necessarily be any conspicuous luminous activity even were it surrounded by a dense stellar system.

One might wonder if there is any evidence for a massive black hole in the middle of our own Galaxy. There is in fact a very peculiar compact radio source right at the galactic centre, which is not like any other kind of known

FIG. 2.14 The radio galaxy Centaurus A. (Photograph from the Hale Observatories.)

radio source: it is not a pulsar, it is not like a supernova remnant, and it is only about a light-hour across. It is a unique source in a unique place, and could be caused by low level accretion on to a large black hole. But this hole cannot be as massive as the one in M87: infrared observations of the neon emission lines from gas near the galactic centre do not show anomalously high speeds; a putative central black hole in our galaxy cannot exceed 5 million solar masses. Our own spiral galaxy can therefore never have been a spectacularly powerful quasar or radio source. It is possible that the most energetic phenomena tend to occur in elliptical galaxies, where the effects of angular momentum are less able to inhibit the accumulation of central gas.

Galaxy formation

A consensus view is emerging that individual galaxies are essentially self-gravitating aggregates where the conversion of primordial gas into stars has occurred or is continuing. In some galaxies, gas accumulates in the centre and, after undergoing a runaway catastrophe, has spectacular consequences manifested in the form of quasars or radio sources.

But let us now broaden our perspective to that of the cosmologist, for whom galaxies are just 'markers' or 'test particles' scattered through space which indicate how the material content of the universe is distributed. When a cosmologist makes a statement about the large-scale smoothness and homogeneity of the Universe, he is referring to the distribution of galaxies; when he speaks of the 'expanding Universe' he is talking about the motions of galaxies.

Galaxies are plainly concentrated in groups of clusters, rather than being distributed purely at random. Our own local group contains the Milky Way and Andromeda galaxies, together with at least 20 smaller members; it is a few million light years across. In fact the Andromeda galaxy is coming towards us at a speed of about 100 km/s. Some clusters of galaxies, such as the one in Coma (Fig. 1.3), contain many more members than our local group. But on a really large scale the Universe genuinely does seem smooth. If one imagined a box whose sides were 100 million light-years (dimensions still small compared to the observable Universe) its content would be the same wherever we placed it in the Universe. In other words, there is a genuine 'broad brush' sense in which the Universe is indeed homogeneous. As one looks at fainter galaxies one probes to greater distances; and the deeper one looks, the less evident clustering is and the smoother the cosmos appears.

How did clustering develop? In fact the simplest possible answer to this question works surprisingly well. If one supposes that galaxies started off being randomly distributed at a time when the Universe was 20 times more compressed than it now is, one can calculate to what extent clumping would subsequently develop due to random fluctuations. The results of such calculations, shown in Fig. 2.15, display a gratifying resemblance to the actual sky. More quantitative statistical arguments do indeed confirm that the clustering of galaxies can be validly generated by this type of process.

But how did individual galaxies form? When the Universe was compressed by a factor of 20, the galaxies themselves would be almost touching. But this is already a late stage in the evolution of the Universe compared to those that Professor Sciama has discussed (it corresponds to a time several hundred million years after the Big Bang). When the entire Universe was in its hot fireball phase, squeezed to densities much higher than their present mean density, galaxies could not have existed in anything like their present form. At those early epochs, the embryonic galaxies could have been no more than regions of slightly enhanced density, whose subsequent expansion was retarded and eventually halted by their excess gravitational field. The main aim of recent theoretical work on galaxy formation has been to try to understand the occurrence of structures with the characteristic observed scales without artificially choosing the initial conditions with this end in view.

The early Universe must have been slightly irregular—otherwise it would still be a uniform gas with no galaxies, no stars, and no observers like

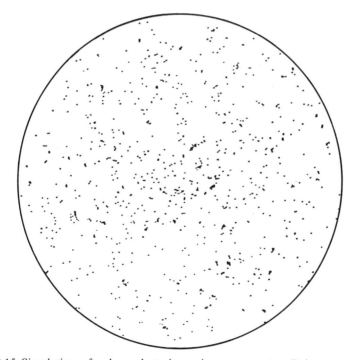

FIG. 2.15 Simulation of galaxy clustering using a computer. Point masses were initially distributed randomly. Clear evidence for clustering has developed at this stage of the calculation.

ourselves to wonder about it. On the other hand, we know the Universe is not completely chaotic, because of the observed large-scale smoothness manifested by the Hubble law and the microwave background radiation. Why it has just this small degree of initial lumpiness is a complete mystery. To the best of my knowledge there are not yet any very convincing suggestions to explain it. One would like to know to what extent the detailed properties of the galaxies are coded or imprinted by the initial conditions. If they all were, then one would never be able to explain the properties of galaxies in any real sense beyond saying that 'things are as they are because they were as they were'. On the other hand, some properties of galaxies, such as their shapes or rotation rates, may be determined by gas dynamics and gravity in a way that the astrophysicist can hope to calculate. Relevant clues may come by comparing the properties of galaxies in clusters with those that are isolated. This will help to distinguish environmental effects from those determined by early conditions.

The subject of galaxy formation and evolution is a kind of bridge between the study of cosmology and the rest of astrophysics. Plainly the problem of galaxy formation must be framed in a cosmological context. Galaxies formed when the Universe was very different from its present state. On the other hand, most astronomical processes proceed regardless of the properties of distant matter and of the cosmological model. I should now like to discuss what galaxies can tell us about cosmology.

Cosmology: the fate of the Universe

Having convinced oneself (or been convinced by Professor Sciama) that the Universe started with a bang, one might ask about its future and eventual fate. Will it continue expanding for ever, so that the galaxies fade and disperse? Or are we in the kind of Universe where the expansion will eventually halt, the galaxies subsequently displaying blue-shifts rather than red-shifts, and being compressed and eventually destroyed in a fireball like that from which they are believed to have emerged? Galaxies would—in this scenario—lose their identity about 100 million years before the 'big crunch'; stars would subsequently be destroyed because the radiation shining on them from the sky would be hotter and brighter than the radiation in their interiors.

The traditional approach to cosmology has been to extend the work of Hubble to study galaxies at greater distances, or equivalently to greater look-back times. The red-shift of a distant galaxy tells us its speed at the time the light set out on its journey towards us. Thus in principle one can learn about the expansion of the Universe at early times, and thereby infer how much it is decelerating. In practice, this type of work is bedevilled by three problems. Firstly, there is an inherent scatter in the properties of the galaxies. Secondly the distant galaxies (which are of greatest interest) appear so faint, even when observed with the largest telescope, that one cannot readily probe to such great depths that the effects of the deceleration would really show up. Thirdly, the galaxies seen at large distances are systematically younger than nearby ones. Even if one can assume that a certain class of galaxy provides a 'standard candle' at the present time, one needs to know how each candle changes as it burns. The galaxies that are crucial for this test are, in effect, being seen at only half their present age: the light has been on its journey towards us for billions of years.

The evolutionary correction required to take these effects into account is at the moment so uncertain that it would be impossible to allow for it and infer the deceleration of the Universe, even if observations *were* extremely precise. There are two aspects to the evolutionary correction. In a younger elliptical galaxy many stars would be shining which by now have died, and the present stars would be seen at an earlier stage in their evolution. This alters the brightness and the colour of the galaxy, the trend being that a younger galaxy would appear somewhat brighter than its present-day counterpart. But there is a secondary evolutionary correction stemming from the fact that one may not be justified in considering a galaxy as a self-contained isolated system. We can see many instances where galaxies seem to be colliding and merging with each other, and in rich clusters such as Coma the large central galaxies may be cannibalizing their smaller neighbours. Fig. 2.16 shows some cases of galaxies in the process of merging. If we were to observe such a system in a few hundred million years it would have settled down into a homogenized single bloated galaxy. Maybe in a few billion years this fate will affect our own Milky Way and the Andromeda galaxy, transforming the local group into a single amorphous elliptical galaxy. Many big galaxies—particularly the so called CD galaxies in the centres of clusters—may indeed be the result of such mergers, traces of disturbance having by now been erased. This process would obviously result in big galaxies having been, on average, fainter in the past. Since there are two corrections, each uncertain but of opposite sign, it is clear that we must understand galactic evolution better before using galaxies for cosmological purposes. Indeed observations must proceed on a broad front, so that solutions to the galactic evolution and cosmological problems will swim into focus simultaneously.

Quasars are hyperluminous beacons enabling us to probe deeper into space, and thus further back into the past, than is possible with normal galaxies: some are so far away that their light set out when the Universe was 20 per cent of its present age. Nevertheless, they have proved no help whatever in geometric cosmology. This is because the evolutionary corrections are even larger and more uncertain than for normal galaxies. The observations imply that there were more bright quasar-type outbursts in the past than now: galaxies must have had a greater propensity to indulge in this type of outburst when they were young than they do now, maybe because there was then more uncondensed gas available for accumulation in the nucleus. Since we cannot yet quantify this correction, we cannot assess

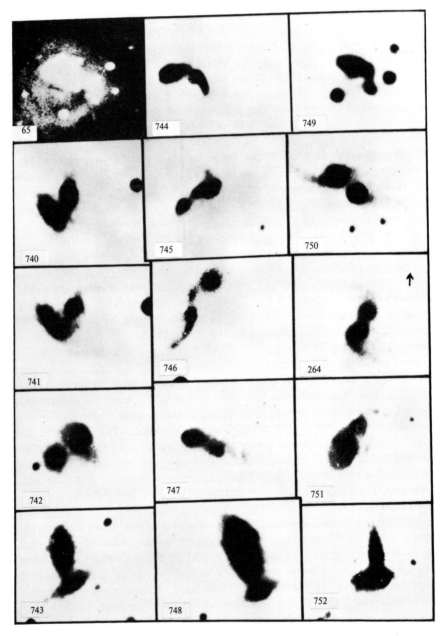

FIG. 2.16 Examples of galaxies in the process of merging. (From *Astronomy and Astrophysics*.)

to what extent the brightness of distant quasars is affected by geometrical effects or the curvature of space.

But there is another more indirect way of trying to determine whether the universal expansion is destined to stop and go into reverse. Imagine that a big sphere is shattered by an explosion, the debris flying off in all directions. Each fragment feels the gravitational pull of all the others and this causes the expansion to decelerate. If the explosion were sufficiently violent, then the debris would fly apart for ever; but if the fragments were not moving quite so fast, gravity might bind them together strongly enough to bring the expansion to a halt. The material would then recollapse. More or less the same argument probably holds for the Universe. One might feel somewhat uneasy about applying a result based on Newton's theory of gravity to the whole Universe. But even though one cannot describe the global properties of the Universe properly, nor the propagation of light, without using a more sophisticated theory such as Einstein's general relativity, the dynamics of the expansion *are* the same as those given by Newton's theory. One can therefore rephrase our earlier question as 'does the Universe have the escape velocity or not?'

In the case of the galaxies (which for the purposes of this argument are regarded as fragments of the expanding Universe) we know the expansion velocity. What we do not know so well is the amount of gravitating matter that tends to break the expansion. It is easy to calculate how much material is needed to halt the expansion: it works out at about 3 atoms per cubic metre. If the average concentration were below this 'critical' density, we would expect the Universe to continue expanding for ever. But if the mean density exceeded the critical density, the Universe would seem destined eventually to recontract.

It is straightforward to estimate how densely the galaxies are packed in the Universe: there turns out to be about one galaxy in every 10^{20} cubic light-years. If one infers the masses of galaxies straightforwardly from the amount of material in the stars one sees, one calculates that the material would provide about only one-thirtieth of the critical density if it were spread smoothly through the Universe.

There is however a difficulty in estimating the masses of galaxies. The straightforward way of doing this is to apply Kepler's law, or a simple generalization of it, to the orbits of stars in a spiral or elliptical galaxy. But if one applies an analogous argument to the orbital motion of two galaxies which seem to be gravitationally bound to each other, or to the motions of galaxies in the cluster, one infers masses almost 10 times higher than those inferred from the internal dynamics. This may mean that the luminous part of every galaxy is embedded in a much larger amount of diffuse dark material (a 'halo') around it. The inference that 90 per cent of the content of the Universe is in some unobserved and unknown form certainly should quench any optimism we may have that galaxies will be quickly understood. On the other hand, maybe one should not be surprised at this inference, because there is no reason to assume that all the stuff in the Universe necessarily shines conspicuously. What the optical astronomer observes may be a small and atypical fraction of what actually exists. The inferred missing mass in galactic halos may consist of defunct stars—the remnants of bright, short-lived stars that caused galaxies to be much brighter when they were young.

If one takes this unseen halo mass into account, it raises our total estimate of the matter-density in the Universe to about 20 per cent of the critical value. But maybe there is more material *between* the clusters of galaxies. This could be in the form of hot gas which never made it into galaxies, and a theorist does not need to have much imagination in order to invent other exotic forms it might take. Our present inventory of the contents of the Universe may still be very biased and incomplete. Some forms of matter (neutrinos or intergalactic black holes for instance) are so elusive that they could provide the critical density without there being any hope that we could detect them: absence of evidence is not evidence of absence. So it is easily conceivable that there is enough material to cause the Universe to recollapse. But *if* the only matter is that which we see, or can already infer from dynamical arguments, then the Universe is destined to undergo the heat-death that was first graphically described by Sir James Jeans. This long-range forecast, disturbing though it may be to agoraphobics, is the most reliable that we can derive from the present data.

Conclusion

Let me in conclusion refocus on individual galaxies. Our knowledge and understanding even of the most normal systems is still very sketchy, basically because of observational limitations. The bright and beautiful photographs of galaxies seen in every popular book give a misleading impression: galaxies are low-surface-brightness objects, barely detectable above the night sky, and only a long exposure photograph reveals them at all. Only recently have we acquired systematic data on the sorts of stars in galaxies of different classes, and on the nature of stellar motions in elliptical galaxies. We still have a rather dim and blurred view of the extragalactic Universe. In 1983 the Space Shuttle is expected to launch the Space Telescope into orbit above the blurring and absorption caused by the Earth's atmosphere. This 100-inch telescope will yield pictures showing details 20 times sharper than those obtainable from the ground. Such pictures should resolve individual stars in many nearby galaxies. The Space Telescope should resolve structure in galaxies at large red-shifts. It may even reveal a genuinely young galaxy and thereby confront current ideas on galaxy formation with empirical test.

The Space Telescope, and other concurrent observational advances expected in the next decade, should offer us a chance to 'firm up' the vague framework of ideas that we now have on galactic structure and evolution.

These advances will need to be complemented by theoretical and interpretative work using the best physics and dynamics, and incorporating the best evidence we can glean on star formation from observations relatively close at hand within our own Milky Way.

In the last few decades we have come to understand the physics of individual stars, what determines their sizes, masses, and brightnesses. The quest for an equivalent understanding of galaxies—to understand why the contents of the Universe are aggregated into these distinctive units—will preoccupy astrophysicists in the next decades. This is the most glaring and basic unsolved problem in astronomy; also, as I have tried to explain in this chapter, a better understanding of normal galaxies may be a prerequisite for progress in settling other questions—the nature of quasars, and the dynamics and fate of the Universe itself.

Bibliography

ARP, H. (1966). *Atlas of peculiar galaxies*, Californian Institute of Technology.

AVERETT, E. H. (ed.) (1974). *Frontiers of Astrophysics*. Harvard University Press.

GINGERICH, O. (ed.) (1975). *New frontiers in astronomy* (reprinted articles from Scientific American). W. H. Freeman, Reading.

HAZARD, C. and MITTON, S. (ed.) (1979). *Active galactic nuclei*, Cambridge University Press.

KAUFMANN, W. J. (1978). *Galaxies and quasars*. W. H. Freeman, Reading.

McCREA, W. H. and REES, M. J. (eds) (1979). Origin and early evolution of galaxies. *Phil. Trans. R. Soc.*

SANDAGE, A. (1961). *Hubble atlas of galaxies*. Carnegie Institution of Washington.

SHIPMAN, H. P. (1976). *Black holes, quasars and the universe*. Houghton Mifflin, Boston.

TAYLER, R. J. (1978). *Galaxies: structure and evolution*. Wyckham, London.

TINSLEY, B. and LARSON, R. (eds) (1977). *Evolution of galaxies and stellar populations*. Yale University Observatory.

3

The origin of the elements

R. J. TAYLER

Introduction

For more than two thousand years man has been interested in the identity of the basic building-blocks of matter. For a time, after the identification of the chemical elements, it appeared that the search was over. It seemed that all matter was composed of a finite number of immutable elements. The later discovery of natural radioactivity of radium and uranium indicated that this idea of elements as immutable could not be completely correct. The subsequent identification of the atomic nucleus, and the realization that the nucleus could be regarded as being made up of protons and neutrons, demonstrated that there were building-blocks more fundamental than the chemical elements. It was soon realized that energy could be released either by fusion of light nuclei or fission of heavy nuclei and that nuclear fusion reactions must release the energy radiated by most stars. Although the question of the identity of the ultimate building-blocks is still unsettled, it is nevertheless true that it remains a good first approximation to regard matter as composed of a mixture of chemical elements whose nuclear structure identifies the element. The composition of matter in the Universe may be changed not only by natural radioactivity but also by nuclear reactions in stars and other astronomical objects. It is then important to ask how the present chemical composition is related to the original composition of the Universe and what was the nature of that original composition. It is with these two questions that this chapter is concerned.

By studying the spectral lines in the optical light and other electromagnetic radiation from stars and gas clouds, astronomers are able to learn something about their chemical composition. The derivation of absolute element abundances from the observed spectra is a very complicated process involving a variety of uncertainties related both to the physical conditions in the system being studied and to the properties of the atoms emitting or absorbing the spectral line. However, if two stars are studied whose principal observed properties are very similar, their relative chemical compositions can be deduced much more accurately than the absolute composition of either. It is possible to obtain better absolute abundances for some hot gas clouds than for stars. In addition to the information obtained from spectra, astronomers know something about the chemical composition of the Earth, Moon, and meteorites from direct analysis. This direct information relates only to surface rocks in the case of the Moon, but also to the atmosphere and ocean in the case of the Earth. Some indirect information about the interior of both the Moon and the Earth can be obtained from seismic studies and geomagnetism.

This work indicates that most stars and gas clouds are almost entirely composed of the two light elements hydrogen and helium. There is no known exception to this statement if attention is restricted to gas clouds and to stars which have only completed a small fraction of their history. In such stars no more than a few per cent by mass of the material is in the form of elements heavier than hydrogen and helium (all such elements are collectively referred to as heavy elements by astronomers). In some cases the amount is a fraction of one hundredth of one per cent. The Earth, Moon, and meteorites have a different chemical composition in that they are relatively deficient in light volatile elements. However, when account is taken of the escape of these elements during planetary formation, it seems likely that they were formed out of the same material as the Sun. In addition the properties of the more massive planets, Jupiter and Saturn, can be most readily understood if their composition is much more nearly solar. Fig. 3.1 shows the estimated overall abundances of solar system material.

From estimates of the ages of stars it is found that the oldest known stars in the Galaxy are those with very low abundances of heavy elements. There is also a correlation of heavy element abundance with position in the Galaxy. A schematic view of the Galaxy from the side is shown in Fig. 3.2. It consists of a flat disc (which is generally believed to make up most of the mass of the Galaxy), a nucleus, and a roughly spherical halo, which contains in particular the large physical groupings of stars known as globular star clusters (Fig. 3.3). The stars with the lowest heavy element content are found in the halo or, if they are not at present in the halo, are moving in such a direction and with such a speed that they could have originated there. These correlations of composition with age and place of origin may be two aspects of the same property and this will be discussed further later.

When the chemical composition of stars was first studied, it seemed possible that almost all stars had the same composition, at least in the surface regions which could be studied. If this were true, the observed surface composition might be regarded as the original chemical composition of the Universe. At the same time it was becoming apparent that the observed light output of stars could only be understood if the energy which they radiated was released in nuclear fusion reactions which changed the chemical composition in their interiors. If the surface compositions of stars did not greatly vary, this would imply that one could not learn much about the nuclear reactions which had taken place in stars by looking at stellar surfaces.

Gamow suggested that the observed chemical composition was not to be regarded as something completely arbitrary, which was put into the Universe at its origin. He pointed out that the observed recession of the galaxies suggested that the Universe had originated in a very dense state and he hypothesized that it was then also extremely hot, thus introducing the Hot Big Bang cosmological theory, which we shall discuss further on p. 56. He hoped that nuclear reactions occurring very shortly after the origin

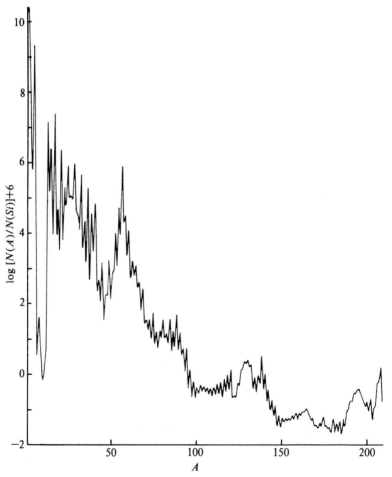

FIG. 3.1 Solar system element abundances. The number of atoms of atomic mass number A is shown on a scale on which the number of silicon atoms is 10^6.

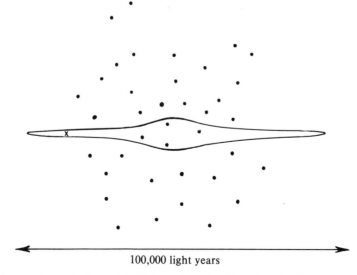

100,000 light years

FIG. 3.2 A schematic view of the Galaxy from the side. Globular clusters are marked
● and the position of the Sun is marked ×.

FIG. 3.3 Globular cluster in Serpens, M5. (Photograph from the Hale
Observatories.)

would produce the observed universal chemical composition. In this he was disappointed as he found that essentially nothing but hydrogen and helium resulted from nuclear reactions occurring in the early Universe. This is mainly a consequence of there being no stable atomic nuclei with mass numbers 5 and 8. Although there were some important errors of detail in Gamow's original calculation, the qualitative character of the results was correct.

This failure to explain the observed abundances in terms of one original event was fortunate as it soon became clear that, although the mixture of heavy elements might not vary much from star to star, the absolute amount can vary by a factor of a thousand. The view was then developed that the chemical composition, essentially purely hydrogen and helium, predicted by Gamow's Big Bang was indeed the original chemical composition of the Universe, but that it had been modified subsequently by nuclear reactions in stars. In order to explain the observed stellar chemical compositions, it is then necessary to assume that stars become unstable towards the end of their life history and return mass to the interstellar medium. New stars can then be formed out of this medium, which has been enriched with heavy elements. This interpretation explains why the heavy element content of young stars is greater than that of old stars. When this suggestion was first made, it was already known that some stars ended their lives by exploding as supernovae with substantial loss of mass.

Towards the end of this chapter I shall discuss the chemical evolution of the Galaxy and will describe how the present composition of the Galaxy can be related to its initial composition by considering what has happened during its life history. A full discussion of this subject is very complex and there are many uncertainties in it at present. The considerations involved in relating the present composition to the original composition include the following:

(1) The rate at which stars form in terms of such parameters as the temperature, density, and chemical composition of the interstellar gas, and the spectrum of masses formed (which may depend on the same parameters).

(2) The extent to which nuclear reactions alter the initial compositions of stars of different masses and the degree of subsequent mixing of nuclear processed material throughout stars.

(3) The amount of mass loss which accompanies stellar evolution, either steadily during normal stellar evolution or catastrophically, possibly at the very end-point of stellar evolution. Our interest is both in the total amount of gas returned to the interstellar medium and in the extent to which its chemical composition has been modified.

(4) The manner in which mass lost by stars is mixed with the ambient interstellar medium and the extent to which we can expect inhomogeneities of chemical composition to be present when a further generation of stars is formed.

(5) The extent to which the Galaxy is a closed system. Does mass lost by stars escape from the Galaxy or alternatively is intergalactic matter accreted by the Galaxy?

Although most of the detailed discussion will refer to our Galaxy, it should be obvious that the same general considerations apply to all galaxies.

It should be clear from what I have said above that a study of the origin of the elements involves knowledge about galaxy formation, star formation, and stellar evolution—together with cosmology if the original chemical composition can be explained in the manner discussed by Gamow. Furthermore the discussion of nuclear reactions taking place in stars requires a wide knowledge of nuclear physics. Much of the nuclear physics was not available until the stimulus from astrophysics excited interest in the particular reactions involved. The same is true of the wide range of atomic parameters needed for the interpretation of spectra.

After this brief introduction I shall turn to each of the major points highlighting as far as possible the major successes and greatest uncertainties. It will be necessary in one short chapter to avoid fine details and to make remarks which, though generally true, are not absolutely correct in the sense that individual exceptions exist.

Observations of chemical composition

Most observations of stellar chemical composition refer to our Galaxy because in most other galaxies stars appear too faint for a detailed analysis of their spectra to be made. We can only study the past chemical composition of a galaxy by analysing the composition of old stars which have not completed much of their evolution. Because the rate of radiation of energy of a star (i.e. its luminosity) depends on a high power of its mass $(L_s \propto M_s^4)$, whereas its nuclear energy supply scales directly with mass, high-mass stars pass rapidly through their life history. In contrast low-mass stars in our Galaxy have not evolved significantly, even if they are as old as the Galaxy itself. Because they are so faint, these low-mass stars can only be studied in our Galaxy (and only in some parts of it). We therefore have information about the past chemical composition of the interstellar medium from which these stars formed for our Galaxy alone. Gas clouds can be studied in more distant galaxies and from these we know something about the present composition of their interstellar medium. In such galaxies there are two types of information which are of interest. The first is the average chemical composition of the interstellar medium compared to that in our Galaxy, and the second is the variation of chemical composition with position in a given galaxy.

Let us first consider the general situation with regard to gas clouds. Galaxies differ in the amount of their mass which is at present gaseous. Elliptical galaxies contain virtually no gas. Galaxies (spiral and irregular) which retain a substantial amount of gas have a tendency for the total heavy element content to be highest in those with lowest fraction of gas. Inside a given galaxy, the fractional gas content tends to be lower near to the centre than further out and there is an anti-correlation of heavy element abundance with gas content in the sense that the heavy element content of a galaxy is higher near the centre than near the edge. From both of these observations it appears that the nuclear (or chemical) evolution of a galaxy is lowest where the gas content is highest. Comparing one galaxy with another, it is not clear whether all galaxies have the same age (as used to be generally believed) or whether some galaxies have formed relatively recently.

The results also strongly suggest that a relation exists between the extent to which galaxies have been enriched in helium and the extent to which they have been enriched in heavy elements. If one uses the standard notation,

X = fraction by mass in form of H,
Y = fraction by mass in form of ^{4}He,
Z = fraction by mass in form of heavier elements,

it is possible to plot Y against Z for a set of gas clouds in a variety of galaxies and to extrapolate the resulting relation to obtain a value of Y when Z was zero. If we believe that the initial composition of the galaxies was *purely* hydrogen and helium, we have then determined the initial helium content. A recent discussion along these lines has given $Y_{\text{initial}} \approx 0.23$ and we shall return to this value later (p. 58).

We now turn to the chemical composition of the stars in our Galaxy and here we can distinguish two types of information. In the disc of the Galaxy lie concentrations of interstellar dust which strongly absorb starlight and as a result it is difficult to observe stars at large distances along the disc (that is along the Milky Way). This means that studies of the chemical composition of disc stars are confined to those which are relatively near to the Sun (or in what we call the solar neighbourhood). In contrast, globular star clusters which lie in the halo of the Galaxy can be studied at large distances because absorption is not a serious problem out of the plane of the galactic disc.

Let us consider observations of globular star clusters (Fig. 3.3). These are known to be amongst the oldest objects in the Galaxy, although our estimates of the ages of globular clusters do not really enable us to decide whether they are all of essentially the same age. Globular star clusters can be divided into those which spend most of the time far out in the halo of the Galaxy and those which are situated much nearer to the galactic centre. All globular star clusters are found to be significantly deficient in heavy elements when compared with the Sun, but the degree of deficiency depends on their position, or more correctly place of origin, in the Galaxy. Those globular clusters furthest from the galactic centre are very deficient in heavy elements whereas many of those nearer to the centre are much less deficient. Although deficient, all globular clusters do contain a measurable amount of heavy elements. If we believe that the Galaxy initially contained no heavy elements, we must ask why the oldest objects of which we have knowledge do contain heavy elements.

Consider next the properties of disc stars in the solar neighbourhood. We should first mention that the solar neighbourhood contains some halo stars, which are at present passing through the disc in their orbits about the galactic centre. At the time that a halo star crosses the disc, it is moving with a much higher speed than stars which are permanently confined to the disc and this enables the halo stars to be identified. These halo stars, like globular cluster stars are also deficient in heavy elements. When these high velocity halo stars are excluded, the evidence for a variation of chemical composition with age in the genuine disc stars is not entirely clear. There is certainly *some* evidence that old disc stars are deficient in heavy elements compared with young stars but in addition (within the accuracy to which stellar ages can be estimated), there is considerable scatter in chemical composition of stars of the same age. The results obtained are shown schematically in Fig. 3.4.

It appears that the heavy element content increased very rapidly early in the history of the galactic disc and there are only relatively few old disc stars which are highly deficient in heavy elements. We shall discuss what these observations may imply later (p. 64).

It is not possible in this one short chapter to discuss variations in the mixture of heavy elements from star to star. Such variations exist and they need to be explained but they are much less important from a cosmological point of view than the variations in the quantity of heavy elements with stellar age and galactic position.

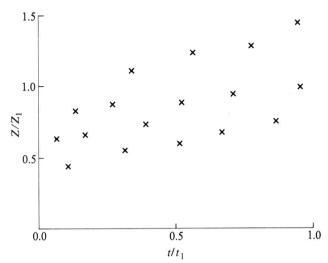

FIG. 3.4 A schematic plot of heavy element abundance versus age for individual disc stars; t_1 is the present time and Z_1 is a typical heavy element abundance of a young star.

A recent development in this field has been the identification of a spectral line due to highly ionized iron in the X-ray emission of gas from clusters of galaxies. Before this discovery was made, it had been generally assumed that intergalactic gas, if it existed, would have the primaeval composition lacking heavy elements. It now appears that some intracluster gas may have a heavy element content comparable with that in galactic discs and this poses a new problem needing explanation.

Turning to solar system abundances we will concentrate on two important points. The first relates to isotopic abundances and the second to heavy radioactive elements. As different isotopes are often produced by different chains of nuclear reactions, it is important to determine isotopic abundances. These can then be compared with the theory described in the next section. When elements possess more than one stable isotope, it is not usually possible to distinguish the spectral lines of the different isotopes in the radiation from stars and gas clouds. The discovery of spectral lines in the microwave region of the electromagnetic spectrum produced by molecules in cold gas clouds has provided a source of information about isotopic abundances of hydrogen, carbon, nitrogen, and oxygen, but for most elements it remains true that isotope abundances are known only for terrestrial, lunar, and meteoritic material. Until a few years ago it appeared that the isotopic abundances of all planetary and meteoritic material were essentially the same, except where alterations had been produced by radioactive decay after the objects solidified. This was presented as an argument supporting a common origin for all solar system material and was used to provide some clues to the processes which fashioned it. This view has been modified slightly by study of one particular meteorite, Allende, in which different fragments of the same meteorite have different isotopic abundances. As a result it has been argued that the interstellar medium out of which the Sun formed could not have been completely chemically homogeneous.

If one studies the present abundances of isotopes of uranium, thorium, and lead in terrestrial and lunar rocks and meteorites, it is possible to deduce how long ago the rocks solidified. This is just one example of the technique of radioactive dating. When applied to this problem it shows that no rocks are older than about 4.5×10^9 years. This is believed to be approximately the age of the solar system. The relative quantities of the three isotopes ^{232}Th, ^{235}U, and ^{238}U now and at the formation of the solar system are very different. But the theories to be mentioned in the next section suggest that nuclear reactions would produce approximately equal amounts of each isotope. The difference between theory and observation must arise because ^{235}U decays more rapidly than ^{238}U, which in turn is more unstable than ^{232}Th. It is then possible to ask when the heavy radioactive elements were formed in order that approximately equal amounts would decay to give the present relative quantities observed in the solar system. It appears that the best agreement between theory and observation is obtained if the elements were produced at a declining rate between about 1.1×10^{10} years ago and the time of formation of the solar system, and this suggests that nucleosynthesis in the galactic disc started about 1.1×10^{10} years ago.

Theoretical considerations

The basic analysis of the nuclear reactions relevant to nucleosynthesis is due to Burbidge, Burbidge, Fowler and Hoyle, and Cameron. There have been some important detailed changes in the subject since their work but only one serious modification of principle. In normal stages of stellar evolution nuclear reactions occur very slowly. In such conditions, if unstable isotopes are produced, they decay before any further reactions occur. The products of a particular series of nuclear reactions then depend on the temperature and density at which the reactions take place but not on the time available. We call reactions like these quasi-static nuclear reactions. In the early discussions of nucleosynthesis, it was assumed that a succession of quasi-static nuclear reactions in stars would be followed by a stellar explosion, which would expel alre ly processed material into the interstellar medium. It was subsequently realized that, when mass loss was initiated by an explosion, which itself involved the sudden raising of stellar material to a very high temperature, rapid nuclear reactions could occur during the explosion. These might significantly modify the nuclear abundances. These explosive nuclear reactions have different end-products from those of quasi-static reactions because they occur so rapidly that they can build on the unstable nuclei which in quasi-static burning would normally decay before undergoing another reaction. Once the importance of explosive reactions had been recognized, it rapidly became clear that it was easier to understand the observed isotopic abundances in the solar system if many of the solar system heavy elements had been created through explosive nuclear reactions. In this case it is important to realize that quasi-static reactions in stars still produce the mixture of elements and isotopes which is subsequently involved in explosive reactions.

Apart from recognition of the importance of explosive reactions, the most significant developments have been a considerable increase in knowledge of the cross-sections for the relevant nuclear reactions together with advances in our knowledge of stellar evolution. These have indicated more accurately the relevant conditions of temperature and density at which the reactions occur.

As described earlier, atomic nuclei can be considered to be composed of neutrons and protons. The mass of an atomic nucleus is less than the mass of the neutrons and protons composing it. This difference in mass manifests itself as energy which is released when a nucleus is formed. The fractional loss of mass is highest for the element iron and its neighbours in the periodic table of the elements. This means that nuclear fusion reactions involving elements lighter than iron can provide energy until the matter has been converted into iron and its neighbours (the so-called 'iron peak elements' because of the local maximum in elemental abundances near iron shown in Fig. 3.1). Given that the elements heavier than hydrogen and helium are rare in stars, there is clearly considerable scope for energy-releasing nuclear reactions which convert hydrogen and helium into heavier elements. Because the rate of nuclear reactions is highly temperature-dependent, and because the insides of stars are very much hotter than their outsides, changes of chemical composition occur in the deep interiors. They can only become apparent to the outside observer if either some mixing process brings the reaction products to the stellar surface or mass loss from the star exposes a processed interior. The fact that stars are rarely discovered with highly unusual surface compositions suggests both that mixing of stellar material to the surface is not, in general, highly important, and that mass loss leading to the expulsion of highly processed material is catastrophic rather than slow, and furthermore that the remnant of such a catastrophe is not a normal star. If slow mass loss were to remove successive layers from a star, there would come a stage when its surface would be made entirely of heavy elements. Such stars are not observed. Some mixing of reaction products to stellar surfaces does occur. This is demonstrated, for example, by the discovery in some stars of the element technetium which has no stable or very long-lived isotope and which cannot have been present when the star was formed. But the occurrence of such anomalies is rare.

Only if interacting nuclei have high relative velocities and are hence at a high temperature can the Coulomb repulsion due to their electric charges be overcome. Thus nuclear fusion reactions involving successively more massive and more highly charged nuclei require successively higher temperatures. The central temperature of a star rises during its early evolution when its material behaves like a perfect classical gas with its pressure being determined by Boyle's law.

The sequence of nuclear reactions which convert hydrogen and helium into heavier elements, culminating in production of iron peak elements, can be classified as follows:

> (i) hydrogen-burning, $1-2 \times 10^7$ K;
> (ii) helium-burning, $1-2 \times 10^8$ K;
> (iii) carbon-burning, 5×10^8 K;
> (iv) oxygen-burning, 10^9 K;
> (v) silicon-burning, $2-4 \times 10^9$ K.

Shown beside the reactions is a characteristic temperature or range of temperatures at which the reactions proceed in quasi-static stages of stellar evolution.

Hydrogen-burning is the most important process because the majority of the energy released by nuclear fusion reactions comes from the conversion of hydrogen into helium. As a result, stars spend most of their active life history converting hydrogen into helium in their central regions, and most observed stars are at this stage of evolution. However, hydrogen-burning produces only helium, which we believe to be already abundant in the initial gas out of which stars were formed.

Hydrogen-burning typically occurs at temperatures between 10^7 K and 2×10^7 K. At the lower temperatures it proceeds through the proton–proton chain, which starts with the conversion of two protons into deuterium and then the addition of a third gives ^{3}He, with several subsequent routes leading to ^{4}He. At higher temperatures it occurs through the carbon–nitrogen cycle in which successive protons are added to what was originally a carbon nucleus until finally ^{4}He and another carbon nucleus are produced.

Helium-burning requires a temperature almost ten times higher than hydrogen-burning. It is a somewhat unusual process. Because of the absence of a stable nucleus of mass number 8, it is not possible for two helium nuclei to fuse. The necessary process involves the conversion of three helium nuclei into ^{12}C. Such three-body interactions are inherently more improbable than two-body interactions. Once ^{12}C has been produced it is possible for a further helium nucleus to be added to give ^{16}O, so that the end-product of helium-burning is a mixture of ^{12}C and ^{16}O.

The remaining reactions listed do not have such simple end-products and, indeed, silicon burning is a shorthand name for a large number of reactions involving many nuclei besides silicon which are the products of carbon and oxygen burning. When the temperature is about 5×10^8 K, it is possible for pairs of ^{12}C nuclei to interact and they produce such nuclei as ^{20}Ne, ^{23}Na, and ^{24}Mg. At a somewhat higher temperature ($\sim 10^9$ K) pairs of ^{16}O nuclei can produce ^{28}Si, ^{31}S, ^{31}P, and ^{32}S. The end-product of carbon and oxygen burning is a mixture of nuclei both heavier and lighter than ^{28}Si, but with ^{28}Si possibly being the most important.

The name silicon-burning has been given to the extensive set of nuclear reactions whereby nuclei in the silicon region are converted to iron peak elements (Cr, Mn, Fe, Co, Ni). These reactions do not involve the fusion of two massive nuclei, because the electrostatic repulsion is now too large. Instead α particles (for example) are broken off one nucleus and added to another, so that as one nucleus is broken down the other is built up. The characteristic temperature for these reactions is a few times 10^9 K. As I have said, the temperatures quoted are those for quasi-static reactions. In an explosive event matter may suddenly be brought to a much higher temperature than the minimum which is necessary for reactions to occur. In that case reactions will take place much more rapidly and produce a different mixture of isotopes. It should also be noted that several different nuclear processes may be occurring at the same time in a star at an advanced stage of evolution and, in the most extreme case, the interior of a highly evolved massive star might appear as in Fig. 3.5.

As the star evolves its central regions become denser as well as hotter and

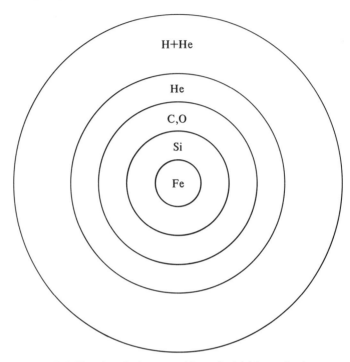

FIG. 3.5 The chemical composition of a highly evolved star.

eventually the matter may become what is called a degenerate gas, in which the Pauli exclusion principle applied to the electrons causes the pressure to depend strongly on the density but to be almost independent of the temperature. At this stage it becomes possible for the temperature to fall while the density and pressure continue to rise. If this happens, the quasi-static nuclear evolution of the star ceases. The temperature at which this happens depends on the mass of the star in the sense that more nuclear evolution occurs in more massive stars. Because of the high dependence of stellar luminosity on mass, massive stars also end their life history more rapidly, as has already been mentioned.

There are some elements which cannot be produced by energy-releasing nuclear reactions. These include all of the elements more massive than those in the iron peak. These are believed to be produced by neutron capture (rapid or slow), the neutrons having themselves been released in fusion reactions, or in some rare cases by proton capture. Finally some rare light elements (Li, Be, B) are thought to be produced mainly by spallation nuclear reactions involving the breakdown of somewhat more massive nuclei by collisions with protons and α particles. The latter reactions are believed to involve cosmic rays and to occur in the interstellar medium rather than in stars. All the elements which cannot be produced by thermonuclear fusion reactions in stars have a low abundance.

An understanding of the nucleosynthesis that occurs in an individual star requires knowledge of the nuclear reaction rates as a function of the relative velocity of the two interacting particles, with the additional possibility that the normal reaction rate might be affected by the environment in which the nuclei find themselves. Thus, for example, at high densities the electrons surrounding a nucleus may be sufficiently close to it to change the effective charge on the nucleus; this is called electron screening. The problems which arise are mainly of two types. In the early stages of stellar evolution the energies of the colliding particles are very low compared with those typically encountered in experimental nuclear physics. This means that observed nuclear cross-sections have to be extrapolated down to very much lower energies, always with the danger that some low-lying resonance, where the cross-section is much higher than is typical, may be missed. The second problem is that, particularly in the case of explosive thermonuclear reactions and rapid neutron capture, many of the nuclei involved are highly unstable so that it is difficult, if not impossible, to measure their properties in the laboratory.

If all of the relevant nuclear data were available, the remaining problems would be solely those of calculating stellar evolution correctly. These can themselves be divided into two basic types. If we consider a star at a slow stage in its evolution and if we specify its mass and its chemical composition we can in principle calculate its structure and hence all of its observed properties. Uncertainties in the results are due to imperfect knowledge of such things as the equation of state of stellar material, the efficiency of energy transport by conduction, convection, and radiation, and the nuclear reaction rates themselves. It is generally believed that these problems are not very important in early stages of stellar evolution, but some doubt has been placed on this view by the results of the solar-neutrino experiment. This experiment was designed to detect neutrinos emitted by the proton–proton chain in the solar interior but the number detected is lower than the number predicted. Until this result is understood, there must be some doubt attached to all of the detailed predictions of stellar evolution theory.

The second type of difficulty arises when we try to relate the structure of a star at one stage in its evolution to that at a later stage or to discuss non-quasi-static stages of evolution. Here the principal computational problems are caused by hydrodynamic motions in stellar interiors, the occurrence of instabilities leading to partial mixing of the stellar material or to mass loss, and to problems during phases when extremely rapid collapse or expansion occurs. Mixing of stellar material may not have much effect on the immediate observable properties of a star, but the redistribution of chemical elements can have a much more important effect on later stages of nuclear burning.

As I have mentioned earlier, there is no evidence for significant numbers of stars whose surfaces are almost entirely composed of heavy elements and this suggests that heavy elements must be returned to the interstellar medium in explosive events. Because the most spectacular stellar explosion is that of a supernova, it is generally believed that the heavy elements in the interstellar medium have been synthesized in stars which became supernovae (Fig. 3.6). There is unfortunately very little direct evidence that supernovae ejecta are rich in heavy elements because, even if they are present, the physical conditions just after an explosion are such that their spectral lines are not prominent. The amount of mass lost in a supernova explosion is also rather uncertain, although it is known to be substantial.

Have there been enough supernovae in past galactic history to produce the heavy elements? The mass of the Galaxy is believed to be between 10^{11} and 2×10^{11} $M_\odot$. ($M_\odot$ is a solar mass, 2×10^{30} kg) and the quantity of heavy elements to be accounted for is probably between 10^9 and

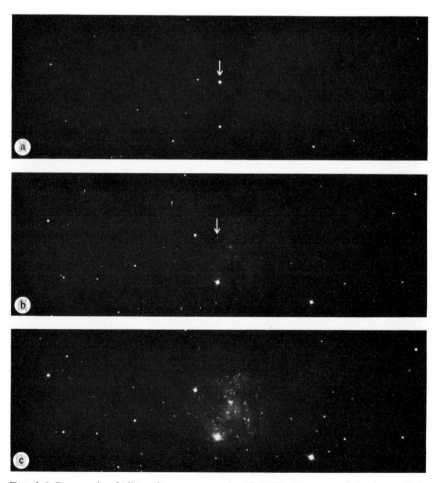

FIG. 3.6 Progressive fading of a supernova in IC 4182. (Photograph from the Hale Observatories.)

$2 \times 10^9 \, M_\odot$. The age of the Galaxy is thought to be between 10^{10} and 1.5 $\times 10^{10}$ years; we have previously given one estimate of 1.1×10^{10} years. We therefore need an average production rate of a little over $0.1 \, M_\odot$ a year. The number of supernovae in the recent past can be estimated in three ways. The three methods involve counts of supernovae in other galaxies (a supernova can be observed in a distant galaxy because at maximum it gives out about as much light as an entire galaxy – Fig. 3.6), counts of shells of matter ejected by explosions (Figs. 3.7, 5.14 and 8.1), which continue to emit radio waves long after the explosion, and counts of pulsars, which are believed to be left behind after a supernova explosion. The results given by these three methods are not in complete agreement but suggest about one supernova every thirty years (possibly as low as fifteen years or as high as fifty). If some allowance is made for an increased rate of stellar birth and stellar death when the Galaxy was young, it seems possible that all of the observed heavy elements can have been generated in this way if the (rather large) amount of about $2M_\odot$ of heavy elements is produced in each supernova explosion.

The question of the extent of mass loss leads us immediately to ask what is left behind when a star explodes. There is evidence for the existence of pulsars (neutron stars) in a small number of supernova remnants. It is not

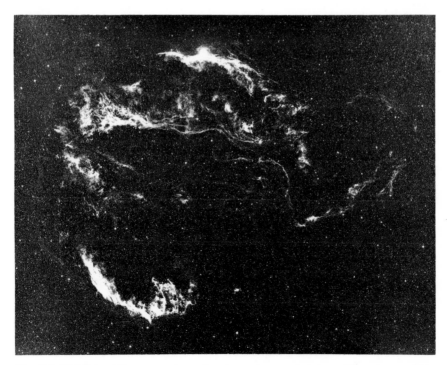

FIG. 3.7 The Cygnus Loop, remnant of a supernova, now forming a filamentary nebula in Cygnus. (Photograph from the Hale Observatories.)

easy to relate the number of pulsars observed in supernova remnants to the number of supernovae which actually produce neutron stars. Even if all supernova remnant neutron stars are pulsars, the number which will be observed will depend on the beaming mechanism of the pulsar radiation. As a result only a fraction of the total number of pulsars are visible. In addition some pulsars are given high velocities by the explosion, so that their identification with a supernova remnant may be difficult. If all supernova remnants were pulsars (which have a maximum mass of about $2\,M_\odot$), the typical mass loss in a supernova explosion would be large enough to account for the heavy elements. It is however possible that some supernovae have more massive remnants than $2\,M_\odot$ which become black holes, and the mass loss would be correspondingly less.

Theoretical calculations of the occurrence and consequences of a supernova explosion have proved extremely difficult. Most calculations have assumed spherical symmetry in the precursor star. However, it is possible that departures from spherical symmetry exist in a highly contracted stellar core just before an explosion and have important effects. Early ideas on supernova explosions suggested that the dynamical collapse of the core of a massive star would be followed by explosive nuclear reactions in material outside, which was suddenly raised to high temperatures. Although it appears that enough energy could be released to disrupt a star, more detailed calculations indicate that this does not in fact happen. It was then suggested that neutrinos, which are emitted in large numbers in a stellar core at late stages of stellar evolution, might be absorbed further out and blow off the outside. At present, it does not seem that this mechanism works either, at least by itself. It is now believed that a major role is played by a shock wave reflected by the compact core. However, it is premature to say that the extent of mass loss and the yield of heavy elements in supernovae is understood.

Cosmological element production

I have already said that according to the Hot Big Bang theory of the Universe the 'initial' chemical composition after the first few minutes is essentially a mixture of hydrogen and helium, with a small but important admixture of deuterium, ^{3}He and ^{7}Li. I have also mentioned observations of gaseous nebulae which suggest that the initial composition by mass of some galaxies in our neighbourhood was 77 per cent H and 23 per cent He. In this section we concern ourselves with the relation between these observed values and the values predicted by the theory.

The theory is suggested by a small number of important observations.

  (i) The spectral lines in distant galaxies are shifted to the red, suggesting that the Universe is expanding.
  (ii) There is an approximately isotropic distribution of galaxies and radio sources.

I. The Orion nebula, probably the most spectacular nebula visible in the sky. The nebula consists of a huge tenuous cloud of hydrogen containing bright young stars, and stars which are still forming.

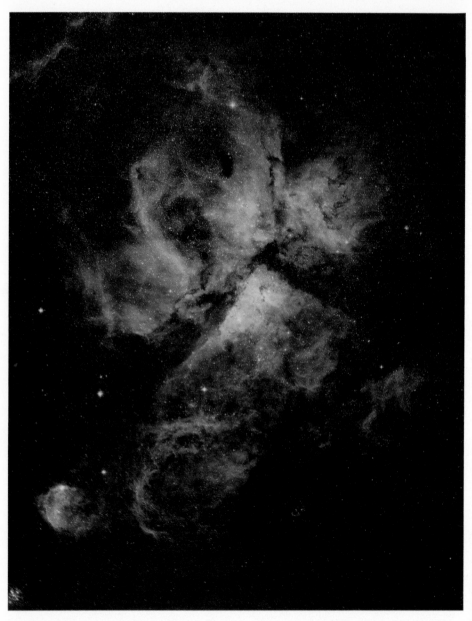

II. The diffuse nebula in the Southern Milky Way known as the 'Key-hole Nebula'. The nebula is partially obscured by dark bands of intervening gas and dust.

(iii) In our locality the number of galaxies is approximately equal in similar large volumes of space.

(iv) Distant radio sources and nearby radio sources have very different properties.

(v) Space is filled with isotropic black body microwave radiation at a temperature of 2.7 K.

(vi) Most objects are made of hydrogen and helium and a very small quantity of heavy elements.

Observations (i) to (iv) suggest that the Universe is homogeneous, isotropic, expanding, and evolving, and these properties are built into the theory. Observations (v) and (vi) are then expected to be consequences of the theory.

The following assumptions are made in the conventional Hot Big Bang theory of the Universe.

(i) The laws of physics are the same at all points and at all times and as a result they are those established in our own locality. It is generally stated that there are four forces of physics:

    (a) gravitation (the General Theory of Relativity),

    (b) electromagnetism,

    (c) strong nuclear interaction,

    (d) weak nuclear interaction,

although current developments in elementary particle physics may lead to a unification of two or all of the last three.

(ii) The Universe is strictly homogeneous and isotropic in its early stages. By this we mean that at a given time the properties of the Universe are the same at all points and the structure is the same in all directions.

(iii) At an initial time $(t=0)$, the density (ρ) and the temperature (T) are both infinite.

With these assumptions, the properties of the Universe are determined by its contents. At very high temperatures elementary particles are to a high degree interconvertible (subject to the requirement that certain quantities such as total electric charge are conserved). For this reason we must specify values of the absolutely conserved quantum numbers of physics. Since at present views about the nature of the interactions between elementary particles are changing, it is uncertain even what these conserved quantum numbers are. Initially we shall assume that there are four:

(i) charge number, C;

(ii) baryon number, B;

(iii) electron lepton number, L_e;

(iv) muon lepton number L_μ;

so that there is in principle a fourfold infinity of Hot Big Bang theories in which all of C, B, L_e and L_μ may be arbitrarily chosen.

The electric force between two charged elementary particles is so much greater than the gravitational force that it is apparent that the net electric charge must be small, otherwise the Universe would be dominated by electricity rather than by gravitation. It is generally assumed that $C = 0$. The baryon number is associated with particles such as the proton and neutron which possess a strong nuclear interaction, and the lepton numbers with particles such as the electron, muon, electron neutrino, and muon neutrino, which interact through the weak interaction but not through the strong interaction. It is also usually assumed that L_μ is zero and that L_e is equal to B. With these assumptions, there remains a single infinity of model Universes to compare with observation.

The baryon number, which can be expressed as the number of baryons per unit volume at some temperature, T, or equivalently as the baryon number density divided by the photon number density, can be related through the theory to the baryon density at the present epoch and to the present temperature (2.7 K) of the black body microwave radiation— believed to be highly red-shifted radiation which has survived from early stages of the Universe. Thus, if we had good knowledge of the present mean density, ρ_0, of the Universe, we should know B. As Professor Sciama has emphasized we do not have a good value for ρ_0. It is certainly more than the smoothed out density of all known matter in galaxies since there is probably a significant amount of matter in other forms. The uncertainty in ρ_0 probably extends about one order of magnitude in each direction about a value which is about $\frac{1}{10}$ that required to close the Universe; that is which would lead to a Universe in which the present expansion will ultimately be replaced by contraction and collapse to a 'big crunch'.

The nucleosynthesis which occurs in the early stages of the Universe depends on the value of B. Although for all plausible values of B there is a split between hydrogen and helium which is qualitatively in agreement with the observations, the actual amount of helium produced increases with increasing B. Within the allowed range of ρ_0, the helium production in the standard theory is between 23 per cent and 27 per cent by mass. If we accept the helium abundance derived from the gaseous nebulae observations referred to on p. 46 and if the standard theory is valid, we must suppose that ρ_0 has close to the minimum value and that the Universe is thoroughly open. The production of D, ^{3}He and ^{7}Li depends much more critically on B than does the production of He. It is not so easy to relate present abundances of these isotopes to their original abundances because within galactic history they are much more easily destroyed. However, because their primaeval abundances vary so strongly with B, it is possible to deduce that, as with the helium abundance, theory and observations are only compatible if ρ_0 has a low value.

The standard theory is not at present in obvious disagreement with any of the relevant observations, although the situation is not completely clear, as I shall explain below. In addition there are some features of the theory which are not universally regarded as satisfactory. One of the worries is that the value of B (defined now as the ratio of baryon density to photon density) required to give the correct element abundances is very small ($\sim 10^{-8}$). Some authors are unhappy that this number is both very small and arbitrary. To overcome this difficulty one suggestion assumes that the Universe was initially very cold—a 'Cold Big Bang'—so that there were initially no photons, and also assumes that both the microwave radiation and the observed photon:baryon ratio were produced in a first generation of stars or more massive objects which preceded the formation of galaxies. There are several problems with this idea. It is not clear that the present large helium abundance can be produced by these first-generation objects. Furthermore there are anyway a number of free parameters in the Cold Big Bang.

A further and stronger objection to the standard theory has been raised by Alfvén and Omnès, for example, who have suggested that the symmetry between matter and antimatter apparent in elementary particle physics should also hold globally in the Universe. Omnès has discussed a cosmology in which $C = B = L_e = L_\mu = 0$. In such a theory there is the problem that the present matter–antimatter asymmetry in our part of the Universe must arise despite the initial symmetrical state. Most people believe that the symmetric theory fails on two points, one theoretical and one observational. The first is that it has not been convincingly shown that large-scale matter–antimatter separation can occur and the second is related to the failure to observe evidence for matter–antimatter annihilation at their common boundaries.

These worries about the Hot Big Bang are in a sense philosophical. There is no empirical need to consider them. A problem however which is not fully resolved relates to the present observed large-scale structure of the Universe in the form of galaxies and clusters of galaxies. In the standard theory complete homogeneity is assumed for the initial state but it now seems that the observed structure cannot arise from a Universe which was initially smooth and homogeneous. Whether or not cosmological element production is affected by such inhomogeneity depends on its extent during the period of element production since regions of higher than average density will have higher helium production.

The helium production depends on how many stable low mass or massless elementary particles there are. It has recently become apparent that there is a charged lepton more massive than the electron and the muon and this has been given the name tauon. It has been conjectured, but not yet proved, that the tauon has a neutrino associated with it and that the neutrino is massless and stable. If it does have these properties, it affects the predicted helium production in the Big Bang which is increased to the approximate range 25 per cent to 29 per cent by mass. If the quoted observational figure of 23 per cent is accurate, the standard model is already in difficulties with this one additional set of neutrinos. Even if we allow for uncertainties in the observed abundances of helium and the other light isotopes, only a very limited number of additional stable, massless neutrinos would be permissible before the predicted abundances would be in conflict with the observations.

I cannot describe here all the interrelations between elementary particle physics and cosmology. It is however necessary to ask whether, if the standard model should be found to be in irrevocable conflict with observations, the other advantages of the Hot Big Bang theory must be sacrificed. The simple answer is no. We have already indicated that the standard theory assumes particular values of the quantum numbers such as L_e and L_μ, and we can consider a theory with different values for these and for the tauon, L_τ (if such a number exists). In particular a reduction in helium production relative to the standard model is achieved if there is a substantial excess of electron neutrinos over antineutrinos. It seems probable that some version of the Hot Big Bang theory will be able to provide an initial element composition for galaxies which is compatible with observations, although there could still be some problems with the abundances of D and ^{3}He, which we have not discussed in detail here. A further problem would arise if a significant variation in the initial composition of galaxies were found. But even that would not necessarily be serious if initial inhomogeneities were needed to produce galaxies and if the observed differences in initial composition are compatible with those which would arise in an inhomogeneous Big Bang.

At present there is no good experimental evidence about either the mass or the stability of neutrinos, although if they were not both massless and stable a serious revision of theoretical ideas concerning elementary particles would be required. Cosmological nucleosynthesis depends on these assumed properties of neutrinos, which means that astronomers watch present developments in elementary particle physics with considerable interest. Another uncertainty relates to the status of the conserved quantum numbers. They are not all conserved in black holes and they may not have been conserved in the very earliest stages of the Universe, before the stages which we have been discussing in this chapter. Indeed there is some hope that the observed value of the baryon number, B, might arise naturally out of the processes of elementary particle physics occuring in the very early Universe.

The ideas described in this section can be summarized as follows. Initially the agreement between theory and observation over the helium abundance in the Universe was taken as good supporting evidence for the Hot Big Bang cosmological theory. From now on it is more likely that it will be assumed that the hydrogen:helium ratio was fixed cosmologically and this ratio will then be used as a constraint on permitted variations of cosmological theory.

Galactic nucleosynthesis

In this section we will assume that the ideas outlined in the previous section are essentially correct and that when galaxies were formed the chemical composition of the Universe was predominantly H and ^{4}He, with a small mixture of D, ^{3}He, and ^{7}Li. What progress has been made in understanding the present composition of galaxies in terms of processes which have occurred since they were formed? Here there are two separate questions. One is concerned with the abundances of particular heavy elements, while the other is concerned with the overall content of heavy elements. In the past most emphasis has been placed on the former question, with an attempt to understand which of the nuclear burning processes would produce a particular heavy element or isotope. Only then did the further question arise as to how the overall combination of heavy elements observed in a particular star could be explained. More recently most emphasis has been placed on the overall heavy element content.

There are several reasons for this change of emphasis. I have already mentioned that the total amount of heavy elements tends to vary much more from star to star than the mixture of heavy elements. This similarity of the elemental mixture in stars of different ages and places of origin was once thought to be a problem because it was believed that supernovae of different types would produce very different mixtures of elements. This should therefore be reflected in the composition of stars containing their ejecta. Although the basic idea is sound, it is now recognized that large numbers of supernovae will have contributed to the heavy elements contained in any given star. In that case the observed approximate uniformity of the mixture of heavy elements might arise because most stars contain material from a representative sample of supernovae.

There are several questions remaining, resulting from the fact that stars of different masses evolve at different rates, so that the earliest heavy elements could not have come from the whole range of supernova masses. In addition mass loss from lower mass stars, which never become supernovae, does have some effect on the abundances of the lighter heavy elements such as carbon and oxygen so long as the galaxy is old enough for such stars to have completed their evolution. It is, however, reasonable to discuss in the first instance the evolution of the chemical composition of a galaxy in terms of the total mass of heavy elements, without considering its detailed composition.

The evolution of a galaxy can be divided into two phases: the initial collapse and formation phase, in which star formation and stellar evolution proceed against a background of very rapidly changing galactic properties, followed by a quasi-static stage in which all galactic properties (such as shape, size, and gas content) change slowly on a time-scale determined by the processes of star formation and stellar evolution. In the case of our Galaxy, it is generally believed that the halo belongs to the initial collapse phase and that the globular star clusters were formed then. As already mentioned, all globular star clusters have a heavy element content which is substantially less than the present heavy element content of the galactic disc.

Superficially there is a problem in that no objects are found devoid of heavy elements; that is with the hypothetical initial composition of the Galaxy. A related problem is that there is a variation of heavy element content in the halo, in the sense that those globular clusters which are confined closest to the centre of the Galaxy are richer in heavy elements than those clusters whose orbits take them far out into the halo. It is possible however that the heavy elements observed in the globular clusters were produced in the earliest stars evolving in the halo itself. If mass loss from these early halo stars then concentrated towards the centre of the Galaxy, this might account for the observed gradient of composition in globular clusters.

As the halo objects have a much lower heavy element abundance than the present galactic disc, it is reasonable to suppose that the disc started with a low heavy element abundance, even if it received some later enrichment from the halo. As a first approximation we take it to be zero. We next ask whether regions at different distances from the galactic centre will have evolved independently or whether the evolution of the whole system is closely linked. There is little evidence for strong gas motions coupling different parts of the galactic disc. In addition the random motions of typical disc stars only cause their distance from the galactic centre to vary by about one kiloparsec (3×10^{19} m) throughout their orbits. We shall therefore start with the assumption that each neighbourhood in the Galaxy evolves independently. We must then ask whether the galactic disc is a closed system: that is, whether a substantial amount of mass escapes from the Galaxy or whether there is significant accretion of mass from the intergalactic medium. If either occurs the disc is not closed. The evidence here is unclear, particularly regarding gain of mass through accretion. We shall start by considering the Galaxy as a closed system, but we shall later reconsider this along with some of our other assumptions.

The chemical evolution of such a closed system depends on the following factors.

(i) What is the rate at which stars form as a function of the properties of the interstellar gas and what is the spectrum of masses formed? In the simplest theories it is assumed that the rate of star formation depends only on the density of the gas and that the spectrum of masses (the so-called 'initial mass function') is also independent of position and time. Both of these assumptions are questionable. The assumption that the star-formation rate depends only on the density of the interstellar gas is certainly wrong in detail, but it is possible that the average rate of star formation over a time which is important in galactic evolution can be expressed as a function of gas density alone.

(ii) What is the yield of heavy elements from a given mass of gas which goes into stars and when are the heavy elements returned to the interstellar gas? With the assumption of constant initial mass function for the stars, the yield of heavy elements can be taken to be independent of time, so long as the total heavy element abundance remains small (as it does to date in all known systems). Obviously the heavy elements are returned to the interstellar medium gradually over a finite time, but with a constant initial mass function the time-scale of return will always be the same. A frequent assumption is the so-called instantaneous recycling approximation. In this it is assumed that the stars which are important in nucleosynthesis are so massive and evolve so fast that the heavy elements which they produce are effectively released immediately (in terms of time-scales which are important in galactic evolution) after the stars are formed. If this assumption is reasonably good, it ensures an approximately constant mixture of heavy elements. This cannot be completely accurate because some nucleosynthesis certainly occurs in low-mass stars with long lifetimes. It is only a matter of detail rather than of principle to allow for non-instantaneous recycling. A further common assumption is that the matter expelled from stars is mixed instantaneously and smoothly into the interstellar medium so that there are no in-homogeneities in chemical composition in a given neighbourhood in a galaxy at a given time.

Before discussing the results, it is of interest to ask what happens in the evolution of different parts of a single galaxy and to compare different galaxies. In a flattened galaxy like our own, the galactic disc will inevitably have a higher density near the centre than near the edge. With the assumption that star-formation depends on the density of the galactic gas to a higher power than unity, gas will be converted into stars more rapidly near the centre than further out. As a result the central regions will evolve more rapidly and we can expect to have radial abundance gradients and variations in fractional gas content. Such gradients are found, though not in all galaxies. The central regions are richer in heavy elements. If we observe instead a group of galaxies, we find that there are differences in overall heavy element abundance and gas content. There are two possible reasons for this. The first is that the circumstances of star formation in individual galaxies are different, so that star formation and nuclear evolution have progressed at different rates. The second possibility is that the galaxies have genuinely different ages. In the past it has generally been assumed that all galaxies have the same age, particularly if we are discussing galaxies in our own neighbourhood, for which detailed chemical compositions can be obtained. This assumption is now being questioned although there is not yet any compelling evidence for more than one epoch of galaxy formation. Even if there are young galaxies, it is to be expected that they formed out of essentially the same primaeval gas with no substantial enrichment in heavy elements.

The simple model of nucleosynthesis outlined above is not successful in describing the solar neighbourhood in our Galaxy. In particular it predicts too many stars with very low heavy-element abundances. Sufficiently low-mass stars, which have formed at any time during the past history of the Galaxy, should be no more than slightly evolved today. Their surface composition should tell us something about the interstellar medium out of which they formed. A study of the number of low-mass stars with different compositions relates the history of heavy element abundance to the star-formation rate. Fig. 3.8 shows that there is indeed a serious discrepancy between theory and observation; the most striking feature is that theory gives a curve which is concave upwards whereas observation gives a curve which is convex upwards.

What is wrong with the very simple picture? It is easy to think of many possible defects including the following:

(i)   the initial mass function may not be unchanging;
(ii) there is significant accretion of gas by the Galaxy and the Galaxy is not a closed system;
(iii) mixing of the galactic gas in the solar neighbourhood is incomplete,

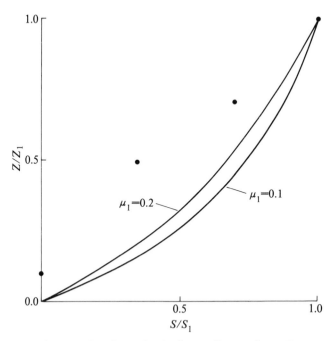

FIG. 3.8 Heavy element abundance in the interstellar medium, Z, at any time is plotted as a function of the total number S, of disc stars formed up to that time. The suffix 1 refers to the present time. μ is the mass fraction in the solar neighbourhood in the form of gas. The solid curves are the predictions of the simple theory for two possible values of the present gas content. The dots are observational points.

so that there are important variations of chemical composition at a given time;

(iv) the instantaneous recycling approximation is inadequate.

It does not seem that the lack of instantaneous recycling does anything more than make detailed alterations in the results. All of the other possibilities have been invoked to try to reconcile theory and observations as described below:

(i) *Prompt initial enrichment.* If the first generation of stars contained only massive stars they would have produced heavy elements, but no low-mass stars would remain today to bear witness to the earliest era with no heavy elements. This is the simplest form of initial enrichment. There must also have been some initial enrichment of the disc due to element production in the halo, but this is not likely to be sufficient to resolve all the difficulties. There have recently been suggestions that our Galaxy has a large low-density massive halo containing matter whose present form is unknown. If there is such a massive halo containing dead stars, heavy elements from these stars may have provided initial enrichment in the galactic disc and, indeed, in the ordinary halo.

(ii) *Infall.* If gas of primaeval composition was accreted by the disc this would clearly have affected the change of chemical composition with time. One effect is to increase the rate of star formation at later stages of galactic history thereby changing the relative numbers of stars with lower and higher heavy element abundance in the direction which is required. Such accreted gas could either be genuine intergalactic gas or be gas which is always in principle part of the Galaxy but which settles slowly to the galactic plane.

(iii) *Metal-enhanced star formation.* Stars are likely to form in places where the galactic gas is coolest, because there gravitational attraction is most likely to overcome pressure due to thermal motions. Heavy elements are more effective at cooling gas than hydrogen and helium. If there are any chemical inhomogeneities in the galactic gas, stars may form more readily where the heavy-element abundance is highest. Stars formed at a given time then have a higher heavy element abundance than the average throughout the gas.

Until recently it was assumed that intergalactic gas would have a primaeval composition with no heavy elements. Now, as we mentioned earlier (p. 48), it is known that gas in some clusters of galaxies contains iron, and presumably other heavy elements. The deduced heavy element abundance is closer to that of the gas presently in galaxies than to gas with primaeval composition. At first sight this is a serious problem. However, it is generally believed that the observation of iron in clusters of galaxies indicates that gas processed in early generations of stars has been expelled from galaxies into the intracluster medium. The very low gas content of elliptical and lenticular galaxies, which are common in galaxy clusters, could indicate that much processed gas does escape from such galaxies into the intergalactic gas.

The discussion of the chemical evolution of galaxies which we have given in this section is very broad brush. At the moment it is not possible to give a unique and definitive discussion of the chemical evolution of our Galaxy, or indeed of any other galaxy. None the less, it is probable that a really satisfactory model of both will emerge in time. Before a full understanding of the subject exists, it will be necessary to understand the origin and distribution of every individual chemical element and indeed of every isotope. It is, of course, necessary to know that there do exist plausible processes which will produce every element. The early work in the subject was more concerned with these detailed considerations. It now seems more profitable to try to understand the evolution of overall composition first, along the lines I have outlined here.

Concluding remarks

In this chapter I have described the progress that has been made in trying to understand the present chemical composition of the Universe in terms of its initial composition and subsequent modifications produced by nuclear reactions during its life history. The idea that the Universe had an origin a little over 10^{10} years ago, in the sense that it has no memory of what, if anything, happened before that, is very attractive. With a hot origin, most of the observed hydrogen and helium can have originated very early in the evolution of the Universe whereas the heavier elements would have been produced by nuclear reactions in stars or possibly more massive objects. It is possible that the simple version of the Hot Big Bang theory will prove unable to account for new observations of chemical composition; or inadequate because of developments in elementary particle physics or because of a failure of the theory to account for new observations unrelated to the origin of the elements. However, at present it is a good working hypothesis against which the observations can be tested, and remains the standard cosmological theory.

Bibliography

Elementary books dealing with cosmology, stellar evolution, nuclear astrophysics, and the origin of the elements.

FOWLER, W. A. (1967). *Nuclear astrophysics.* American Philosophical Society, Philadelphia.
SCIAMA, D. W. (1971). *Modern cosmology.* Cambridge University Press.
TAYLER, R. J. (1978). *Galaxies: structure and evolution.* Wykeham, London.
—— (1972). *The origin of the chemical elements.* Wykeham, London.
—— (1970). *The stars: their structure and evolution.* Wykeham, London.
WEINBERG, S. (1977). *The first three minutes.* André Deutsch (hardback), Fontana (paperback), London.

More advanced texts

CLAYTON, D. D. (1968). *Principles of stellar evolution and nucleosynthesis.* McGraw Hill, New York.

PEEBLES, P. J. E. (1972). *Physical cosmology.* Princeton University Press.

Research papers which have had a major influence on the development of the subject.

ALPHER, R. A., BETHE, H. A., and GAMOW, G. (1948). *Phys. Rev.*, **73**, 803.

BURBIDGE, E. M., BURBIDGE, G. R., FOWLER, W. A., and HOYLE, F. (1957). *Rev. mod. Phys.*, **29**, 547.

CAMERON, A. G. W. (1957). *Chalk River Report*, CRL 41, Atomic Energy of Canada Ltd.

Recent developments relating particularly to the chemical evolution of galaxies will be found only in specialist astronomical journals.

4

The stars as suns

D. E. BLACKWELL

The stellar universe is a relatively quiescent, low-temperature, low-energy universe. Furthermore, our view is rather parochial because ordinary stars, unlike galaxies and quasars, are not very bright candles and cannot usually be seen at very great distances. As a result we shall be concerned in this chapter only with the immediate neighbourhood of the Sun, out to a distance of a few hundred light-years.

Stars are deceptively simple systems. Some years ago an astrophysicist began his lecture by saying that a star is a 'pretty simple thing', only to be heckled by a member of the audience who cried out, '*You* would look pretty simple at a distance of one hundred light-years'. But, putting aside the complexities of real stars, an ordinary star is basically a giant gaseous sphere with a diameter of between half a million miles and 500 hundred million miles. The temperature in the central region is typically a few tens of millions of degrees, whilst the temperature of the surface layers is commonly between 2500 and 20 000 K. It is composed chiefly of hydrogen, with some helium and traces of almost all the other elements. The presence of most of the elements in the Sun and stars is easily demonstrated by spectrum analysis. The spectra of Fig. 4.1 show an example of this. The upper spectrum is of the Sun at a wavelength near 512.8 nm, showing characteristic absorption lines produced by various elements, whilst the lower spectrum is of a laboratory furnace used in absorption and containing iron. The line at 512.769 nm in the furnace spectrum is due to iron. The presence of a line at this wavelength in the solar spectrum shows that iron is a constituent of the solar atmosphere. Most elements can be detected in the Sun and stars by means such as this, and even their relative proportions determined. Studies of this kind show that the majority of stars are undistinguished, in the sense that they contain the various elements in roughly the same relative proportions, they are stable, and they do not possess a large magnetic field. A few stars do have extraordinary properties. The star 53 Tau for example has an

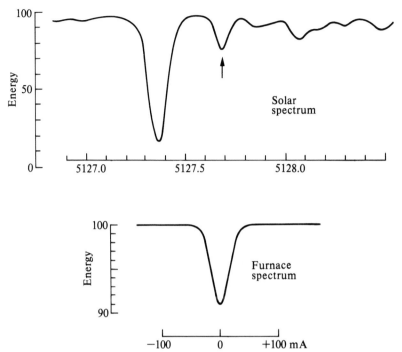

FIG. 4.1 Spectra of the Sun (*upper*) and laboratory furnace containing iron (*lower*). The 512.769 nm line of iron in the lower spectrum is present in the solar spectrum.

enormous excess of manganese with rather less iron than usual. Another star will have a magnetic field of 30 000 gauss. Still others are composed entirely of helium or are without metals.

Here we shall discuss some of the properties of ordinary undistinguished stars, and relate these to current ideas about the way in which stars form and the earlier phases of their evolution. The development of a star will be taken up to the stage in its lifetime when it is beginning to be middle-aged. Indeed the analogy with human beings is a close one because, as we shall see, stars at this period in their lives begin to increase in girth. In even later life some stars become explosively unstable, turning into supernovae, which for a short time may be as bright as a whole galaxy; but that phase will be left to Professor Tayler's discussion of the origin of the elements.

It has been realized since the time of Galileo that stars are very numerous. This is demonstrated by the photograph in Fig. 4.2, which is of a small region of the sky, scarcely bigger than the full moon, in the constellation Sagittarius, taken with the new UK 48-inch Schmidt telescope in Australia. The faintest stars detectable on this photograph have a magnitude of about 23^m. The energy flux from these stars is only about 10^{-10} times the flux from the bright star Sirius. Because of the large amount of information stored in them, photographs like this one are now usually measured automatically

FIG. 4.2 Photograph of a dense star field, about 0.5 degree square, in Sagittarius taken with the UK Schmidt telescope in Australia. (Courtesy Royal Observatory, Edinburgh.)

using a machine such as the one called COSMOS developed at the Royal Observatory Edinburgh, or the machine being developed by Kibblewhite and others at Cambridge. Fig. 4.3 shows the result of an examination by COSMOS of a small portion of a photograph like that of Fig. 4.2, which is 3.3 arcmin square (that is, roughly one per cent of the area of the full moon). The large feature in the upper left is a bright star with a halation ring and spikes produced by the diffraction of light at various supports in the telescope. The 150-inch Anglo-Australian telescope shows fainter stars still, reaching to about 25^m. The 2.4-metre (95-inch) Space Telescope to be launched by NASA should reach even fainter stars using photoelectric detection, perhaps to a limit of about 27^m in an integration time of 2000 s. The Space Telescope will not be gathering more light than a ground-based telescope, but it will make better use of it because it will concentrate the light of each star into a smaller image. But even with the Space Telescope we could not begin to count the stars in the Galaxy. This is mainly because the regions towards the centre and beyond are wholly obscured by interstellar dust. This dust is distributed patchily throughout the Galaxy with a particle density of a few particles per cubic km, and its effects are shown strikingly in Fig. 4.4 and Plates 1, 2 and 3.

Despite this obscuration we do know roughly how many stars there are in the Galaxy. Just as it is possible to obtain the mass of the Sun from the

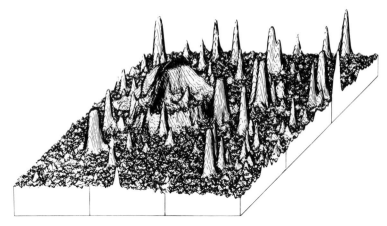

FIG. 4.3 COSMOS map of a 3.3 arcmin square star field. (Courtesy Royal Observatory, Edinburgh.)

FIG. 4.4 Obscuration by interstellar dust in the region of Orion. (Photograph from the Hale Observatories.)

orbital period of the Earth and its distance from the Sun, so it is possible to determine the mass of the central parts of the Galaxy from the orbital period of the Sun around these inner regions. The result is about 10^{11} solar masses, and since the Sun's mass is about average for all stars this number is roughly the number of stars in the Galaxy: about twenty for each man, woman, and child on the Earth. Only a very small fraction of these stars could be photographed, and of these only a few of the brightest could ever be studied individually. The fainter stars mostly have a low surface-temperature and a low mass: stars that have been called by Professor Jesse Greenstein 'the petit-bourgeois members of the lower middle-class inhabitants of our Galaxy'.

As well as dust, the space between the stars also contains a tenuous gas, chiefly hydrogen, from which it is believed the stars originally condensed. The existence of the calcium component of this interstellar gas is shown by the spectra of distant stars in Fig. 4.5. In these spectra, the interstellar line can be distinguished from the line due to absorption in the stellar atmospheres. Interstellar space is so cold, and the gas density is so low, that the interstellar line is sharp, whereas the line due to the star is wide. A substantial proportion of the material of the Galaxy is now in the form of stars rather than a gaseous interstellar medium, so it seems likely that the condensation process is quite efficient.

The details of the way in which the gas condenses into stars are still unclear, but we might expect at first one or a few very large cold protostars to be formed in an initially condensing cloud. These would radiate most strongly in the infrared region of the spectrum and perhaps hardly at all in the visible region: like a poker that is not quite red-hot. Astronomers believe stars are being formed in this way in our Galaxy today, since some stars are so bright that they could scarcely have been born much more than a million years ago—and a time-interval of a million years is negligible in comparison with the age of the Galaxy. So it is relevant to enquire where these embryo

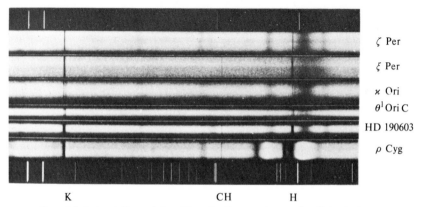

FIG. 4.5 Interstellar calcium lines in the spectrum of a distant star.

stars are to be found. The Orion nebula, shown in Plate 1, is a possible centre of star formation, and there is indeed one infrared source in it discovered by Becklin and Neugebauer with no optical counterpart, which might be such a protostar or group of protostars. The source has a diameter of less than 2 arcsec and a temperature of 530 K derived from a measurement of its flux distribution out to a wavelength of 20 μm.

It is possible that there are even more conspicuous protostars to be detected in the celebrated dark globules discovered by Barnard and more recently studied by Bok. The photograph of Fig. 4.6 shows two such dark areas, and that of Fig. 4.7 shows one of these enlarged. The extreme blackness of the globule and the absence of stars within it is very striking, as is also its sharp boundary. These globules are possibly large clouds of molecular hydrogen and other molecular compounds mixed with dust. The dust absorbs light very strongly so that they appear opaque at the centre. As a result few, if any, stars can be seen through them. However, if they contain protostars they must still be very cool because no infrared continuum flux has yet been detected from any of these globules.

As the hypothetical protostar contracts, the gravitational energy released causes its central temperature to rise until it is hot enough for nuclear reactions to begin; in the first of these reactions, hydrogen is converted to helium with the release of energy. Gravitational contraction then halts. At the same time, the surface temperature of the protostar rises until it is

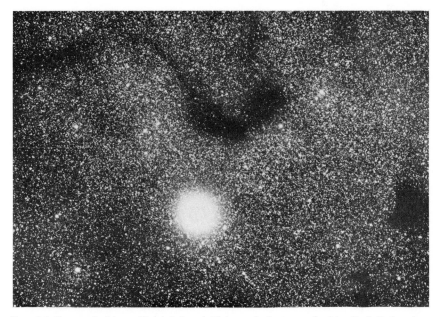

FIG. 4.6 Barnard objects 68 (*right*) and 72 (*upper*) photographed by B. J. Bok using the 4 m CTIO telescope. (Courtesy of the Astronomical Society of the Pacific.)

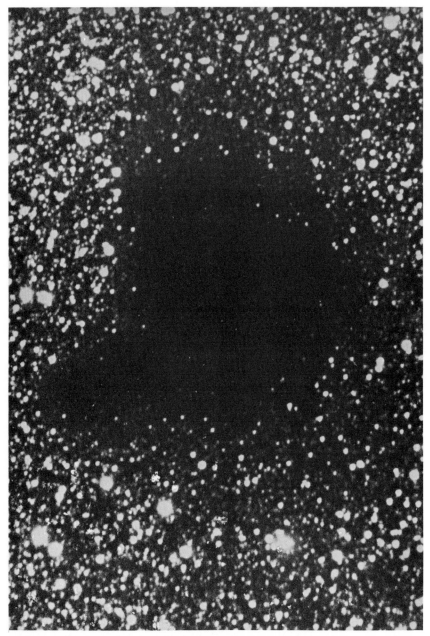

Fig. 4.7 Enlargement of photograph of Barnard object 68 obtained by B. J. Bok with the 4 m CTIO telescope. (Courtesy of the Astronomical Society of the Pacific.)

radiating strongly in the visible and becomes the kind of star that we are familiar with. At this stage, when nuclear reactions start to generate energy, the star is said to have age zero. From this moment onwards the star begins to evolve as a result of changes in chemical composition produced by the nuclear 'burning' reactions. On the whole it does this by getting bigger and brighter. The brighter, more massive stars, evolve quite quickly, on a time-scale of a few million years, but stars like the Sun evolve more slowly. More than 5000 million years must elapse before the earth is engulfed by the Sun. Some stars that are much less massive than the Sun never manage to start nuclear reactions at all. They are destined to live out their dull lives like red-hot pokers, slowly cooling off.

Although the theory of energy generation in stars by nuclear reactions is generally accepted, there is one outstanding difficulty. This difficulty has been raised by a celebrated experiment, started nearly ten years ago, to test the present theory of nuclear-energy generation in the Sun. The basis of the experiment is that when hydrogen is converted to helium, large numbers of particles called neutrinos are generated. A remarkable property of these particles is that they hardly interact at all with their surroundings and are therefore able to pass to the surface of the Sun and escape without difficulty. This is in contrast with the photons of light which are bounced about from atom to atom on their way to the surface and take about a million years for the journey from the centre to the surface. The expected flux of neutrinos at the earth is easily calculated, and is about $6 \times 10^{10} \, \text{s}^{-1} \, \text{cm}^{-2}$. This corresponds to quite a large energy flow, amounting to about 7 per cent of the total energy flow from the Sun. If these neutrinos could be detected then there would be direct evidence for nuclear reactions occurring in the Sun. However, in spite of the large neutrino flux, their detection is far from easy. An experiment to detect solar neutrinos was started by Davis in 1970, using as a detector a tank containing 400 000 litres of dry-cleaning fluid, a compound of chlorine. The tank was buried at a depth of 1500 m in a mineshaft in Colorado. A small proportion of the neutrinos that pass through the tank will interact with the isotope of chlorine, ^{37}Cl, to produce argon atoms. The expectation is that there will be only a few atoms of argon formed per week. None the less, the experimenters are confident of being able to count such a small number accurately. The results obtained over a period of a few years are displayed in Fig. 4.8. From this diagram it will be seen that although there is a large fluctuation in the measured values over the years, the average detection rate is much less than expected. This apparently small flux of neutrinos has been a major setback for studies of the solar interior and there is still no accepted explanation for it.

If we ignore the difficulty of explaining the apparently small solar neutrino flux, theoretical predictions of the way in which a star starts to evolve, beginning from age zero, are well established. The fundamental parameters of a star are its mass, radius, and effective temperature (which we consider more fully later, but can be considered to be roughly the temperature of the surface layers). To these can be added its luminosity, which is the total rate of emission of energy by the star, measured in watts. This includes contributions across the whole electromagnetic spectrum ranging from X-rays through the ultraviolet, visible, and infrared regions. Longer wavelength radio radiation is rarely observed from stars, and then generally the circumstances of its emission are exceptional. Although it is important, the luminosity is not an independent parameter because it is related to the temperature of the star (which determines its rate of emission of energy from unit area) and its surface area or radius.

Stellar evolution theory predicts the way in which the radius, temperature, and luminosity vary with time. The nature of these changes depends principally on the mass of the star and to a certain extent on its chemical composition. The results of calculations by I. Iben using large computers are given in Fig. 4.9, in which the lines show the way in which the calculated luminosity and temperature change with time for a range of stellar masses. The diagram shows that as the less massive stars evolve they first decrease slightly in temperature and then increase sharply in luminosity. This increase in luminosity is chiefly due to an increase in radius, so that radiation comes from a larger surface-area. The more massive stars evolve faster than the less massive ones. For example, the star of mass $15\,M_\odot$ moves to the end of the part of its track given in the diagram in a time of 1.2×10^7 years, but the $1\,M_\odot$ star takes 1.1×10^{10} years to move to the end of its track.

It is clear that important information about the state of evolution of a star, including its age, could be derived from such a diagram if only the

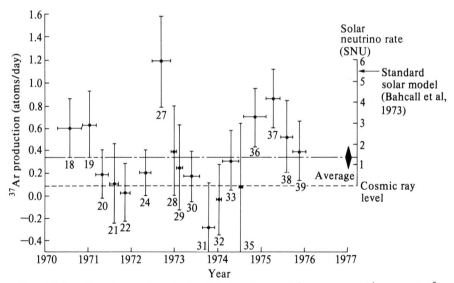

FIG. 4.8 Results of experiments to detect solar neutrinos, expressed as rate of detection of ^{37}Ar atoms per day. The expected solar rate is shown at the right-hand side. (Courtesy of the Astronomical Society of the Pacific.)

parameters of mass, radius, temperature, and luminosity were known with good accuracy. The sort of accuracy astronomers now aim to achieve is 2 per cent in radius, 1 per cent in temperature, and 5 per cent in luminosity. Accurate determination of these parameters is crucial to basic stellar evolution theory, and as it is now a lively and exceptionally difficult area of observational astrophysics, we will consider how the measurements are being made, and plans for the future. The accurate determination of stellar mass, particularly if the star is isolated in space, that is, if it is not a member of a binary system, is a specially difficult problem which will not be considered here.

We begin by discussing the concept of the effective temperature of a star. It is not clear at first sight how a unique temperature can be assigned to a star when its temperature varies so much between its centre and the surface. However, the most significant defining parameter of a stellar atmosphere is the total flux of radiation emitted by unit area, $\mathscr{F}_s$ measured in watts m^{-2}. Because of this, it would be more natural to classify stars by their emitted flux than by temperature. For example, a hot star with a surface temperature of 28 000 °C would emit 3.5×10^{10} watts m^{-2}, whilst a cooler one at 5000 °C would emit 3.5×10^7 watts m^{-2}. But such a system of classification would not be popular among astrophysicists because fluxes cannot easily be

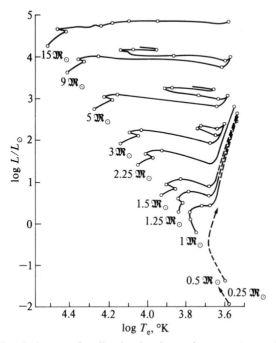

FIG. 4.9 Calculated change of stellar luminosity and temperature with time for a range of stellar masses from 0.5 M$_\odot$ to 15 M$_\odot$.

visualized. It would be like telling a blast-furnace operator that his steel is emitting a total flux of radiation of 7.5×10^4 watts m^{-2} instead of saying that its temperature is 800 °C. Solely for convenience, because temperatures are more easily visualized than fluxes, astrophysicists convert these fluxes to temperatures called effective temperatures, T_e. These effective temperatures are defined by the Stefan–Boltzmann relation $\mathscr{F}_s = \sigma T_e^4$, which relates surface flux per unit area and temperature through the constant σ. In spite of the general use of this concept of temperature, the surface flux from a star remains the fundamental parameter.

It follows from this that the only proper method for determining the effective temperature of a star is to work through this definition and determine the total flux per unit area, $\mathscr{F}_s$. This can be done by measuring the total flux from the star at the earth, $\mathscr{F}_E$, and the angular diameter, θ. Then a simple transformation shows that the surface flux at the star is related to the enrgy flux per unit area at the earth through the angular diameter of the star, θ, as

$$\mathscr{F}_s = 4\mathscr{F}_E/\theta^2$$

and from the relation

$$\mathscr{F}_s = \sigma T_e^4$$

we obtain T_e in terms of observable quantities,

$$T_e = \sqrt[4]{\frac{\mathscr{F}_s}{\sigma}} = \sqrt[4]{\frac{4\mathscr{F}_E}{\theta^2 \sigma}}$$

The first step in determining T_e for a star is therefore to measure the angular diameter, θ. This is extremely difficult, even for the nearest stars, especially as there is little point for this particular problem in obtaining measures that are less accurate than 2 per cent. The magnitude of the difficulty in measuring θ to this kind of accuracy may be illustrated by two examples. Betelgeuse, a nearby star of radius about 600 solar radii, has an angular diameter of only 0.041 arcsec. This is the largest stellar disc known, yet it corresponds to the diameter of a pinhead seen at a distance of 5 km. The star Sirius, which is also nearby but has a radius of only about 2 solar radii, has an angular diameter of 0.0062 arcsec, corresponding to a pinhead at 30 km. Small though these diameters are, we need in fact to determine angular diameters for stars more distant than Betelgeuse or Sirius by a factor of 10 or 100, whilst preserving an accuracy of 2 per cent.

The practical limit of resolution of a perfect telescope of diameter D at wavelength λ is set by the diffraction of light, at $1.22\,\lambda/D$ radians. For the 200-inch telescope this limit is 0.025 arcsec. In perfect conditions, this telescope could therefore just resolve the apparent disc of Betelgeuse, but the effect of atmospheric turbulence (commonly called 'seeing') is to degrade the image quality to a diameter that might be as much as 3 or 4 arcsec, so that it would be quite impossible for the telescope when used conventionally to resolve the disc of Betelgeuse. The Space Telescope will not suffer from this image degradation, but its diffraction limit will be only about 0.1 arcsec, or perhaps sometimes 0.08 arcsec, so that it will not be able to resolve the disc of any star. However, in 1970 the French astronomer Labeyrie showed that the image of a star when photographed with a very short exposure time, using an image intensifier, shows a speckle pattern in which each speckle is an image of the star undegraded by atmospheric seeing and limited in size by diffraction in the telescope. A typical pattern for an unresolved star is shown in Fig. 4.10b. In contrast with this, the speckles in the pattern obtained for Betelgeuse shown in Fig. 4.10a clearly show an increase in size due to the resolved disc of this star. The photographs of Fig. 4.10 are of fundamental importance because they demonstrate for the first time in an obvious and unsophisticated way that one star at least has a disc of finite size. We shall see later that other evidence of a more indirect nature has led in the past to the same conclusion, but the simple and direct demonstration by speckle techniques is most satisfying.

The speckle technique is very promising, with a great deal of development before it. The claimed internal accuracy is good. This is illustrated by the recent determinations by Wilkerson and Worden of the angular diameter of Betelgeuse. They give a diameter of 52 ± 1.7 milliarcsec in good confirmation of the independent result of 49 ± 1 milliarcsec obtained by Lynds and co-workers. The method suffers from the limitation that it is applicable only to stars that show large discs. For Arcturus, for example, which has a diameter of about 200 milliarcsec, the accuracy of the speckle measurement by Worden is only 31 per cent. However, the method is likely to be useful

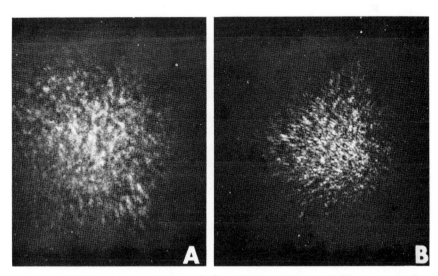

FIG. 4.10 Speckle patterns photographed at Kitt Peak National Observatory using the 4 m telescope (a) α Ori (Betelgeuse) showing evidence of a stellar disc, (b) β Ori, a point source star. (Courtesy of *Vistas in Astronomy*.)

for other purposes. For the fundamental kind of work that we are discussing it is essential to know that the star under study is single and not part of a binary system, and this information can often be provided by the speckle technique. Indeed, the technique is going to be important in studies of binary stars and their orbits. An instance of this is the star 61 Ori which has recently been found to be a binary with a separation of only 0.1 arcsec, using the Anglo-Australian telescope. Such a measurement would be quite impossible by conventional means. The method will probably also be useful for non-stellar infrared sources, but as one cannot take photographs in the longer wavelength infrared region, a scanning system of the kind suggested by Selby and his co-workers will have to be used.

Another direct method for determining θ that is simple in principle involves the occultation of a star by the Moon. The Moon moves against the star background at the rate of about 0.5 arcsec per second of time. Hence the time taken by the moon to pass over the disc of the star 31 Leo, for instance, which has an angular diameter of about 4 milliarcsec, is about 8 millisec (the exact duration of the occultation depending on the position of the star on the Moon's limb). But unless the angular diameter is very large, the situation is not quite as simple as this because the light from the star is diffracted by the limb of the Moon to form a pattern of alternate bright and dark fringes, which sweep across the telescope as the Moon moves. Some uncertainty is introduced through lack of knowledge of the fine structure of the Moon's limb at the place of occultation, but at least the method is independent of any degradation of the image quality by the Earth's atmosphere. The method is now being used in a routine fashion with a best claimed accuracy of 2 per cent, and many published values for stars of large angular diameter, at least, are probably within 10 per cent of the true value. However, like the speckle technique, the occultation method is only suitable for stars that have a large angular diameter.

The most famous apparatus for determining a stellar angular diameter, and one that is described in almost all textbooks of optics and astronomy, is the stellar interferometer of Michelson. The idea was actually conceived by Fizeau in 1868 and taken up by Stephan who built an instrument, observed interference fringes, and deduced that stars have angular diameters that are less than 158 milliarcsec. Michelson started work on the problem in 1890 but did not proceed vigorously with it until some years later because he was unaware until then of the existence of giant stars (that is, stars that have become large through evolution) which have a large angular diameter if nearby. He was also apprehensive about the effect of atmospheric seeing on the measurement of small angular diameters. The interferometer had two mirrors mounted on slides on a long girder placed across the top of the telescope. In operation the observer varied the separation of the mirrors so that the interference fringes, which are superimposed on the image of the star, alternately appeared and disappeared. Fig. 4.11 is a sketch made by Michelson's colleague, Pease, in his notebook to show the night assistant how he should sit balanced precariously at the end of the telescope in order to move the mirrors. Michelson and Pease first used the interferometer to measure the star Betelgeuse, but in spite of their success then, only a few stars were subsequently measured and the method proved to be of limited use. It needed great skill and was dependent on atmospheric seeing conditions; and a bigger instrument built later by Pease gave a disappointing performance. However, the work was of great importance because it provided the first, albeit somewhat indirect, evidence for the existence of resolved stellar discs. The angular diameter of Betelgeuse finally determined by Pease was 41 milliarcsec.

Following the limited success of the Michelson interferometer, various other interferometers have been proposed and tried, but the first spectacular advance was achieved by Hanbury Brown with his intensity interferometer. In this instrument two large steerable parabolic mirrors direct light from a star onto two photomultipliers, which detect the light (Fig. 4.12). In developing the theory of the method Hanbury Brown showed that the correlation between the fluctuating outputs of the two photomultipliers depends on the separation of the mirrors and the angular diameter of the star being observed. In use, the mirrors are mounted on a railway track so that their distance apart can be varied, and during observations of a particular star the correlation between the outputs, measured over a period of a few hours, is obtained as a function of the separation of the mirrors. This novel instrument has given the first stellar diameters of good accuracy,

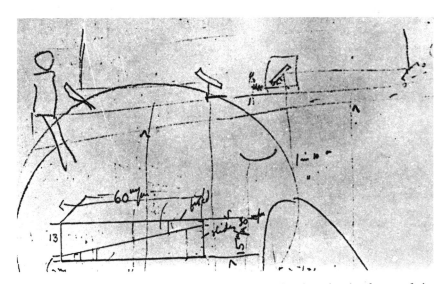

FIG. 4.11 Extract from notebook of F. G. Pease showing sketch of part of the Michelson stellar interferometer and illustrating the precarious position of the night assistant whose duty it was to move the interferometer mirrors. (Courtesy of the *Journal for the History of Astronomy*.)

but it has its limitations. The present design is useful only for bright stars ($V < 2^{m}\!.8$) which must be hot ($T_{c} > 7400\ °C$). It has been used to measure a total of 32 stars, but unlike the Michelson interferometer it is scarcely, if at all, dependent on seeing and does not require a star to show a large disc. As an illustration of this, it has been used to measure the star ζ Pup, which has an angular diameter of only 0.42 milliarcsec, although the seeing may have

FIG. 4.12 One of the multicomponent mirrors of the intensity interferometer. (Courtesy Chatterton Astronomy Department, Sydney, NSW, Australia.)

been as much as ten thousand times as large. The accuracy depends on the brightness of the star, and for $V = 2^m$ the claimed accuracy is about 8 per cent. But even this instrument, fine though it is in concept, scarcely begins to satisfy the needs of astrophysicists, for its rate of measurement is slow, it cannot measure stars that are either faint or cool, and even for bright stars its accuracy, although good, is still not quite good enough. There are plans for a new larger instrument of improved performance which will reach fainter stars but the future of this project is uncertain.

The ideal method for measuring stellar angular diameters should be applicable to all stars irrespective of their temperature and virtually independent of their brightness, and give an accuracy that is of the order of 1 or 2 per cent independently of the actual angular diameter. It seems likely that a photometric method, called the infrared flux method, will satisfy these requirements. The method needs two measurements to be made for a star; these are, the total flux from it, $\mathscr{F}_E$ and the flux at a particular wavelength in the infrared (which might be between 3 μm and 5 μm), both measured at the earth. These two measurements are combined to give both the angular diameter of the star and its effective temperature. A comparison of the results obtained with this method and those obtained using the intensity interferometer is encouraging because for a total of seven stars the difference between the two methods is only 3.8 per cent, but it is not yet known which of the two is giving the more accurate results.

A comparison between the angular diameters obtained for Betelgeuse using the Michelson method, the speckle method and the infrared flux method is instructive and of great interest. The dependence on wavelength of the values obtained is shown in Fig. 4.13. A remarkable feature of the speckle observations is the apparent decrease of diameter with wavelength, which is confirmed by the infrared flux method. It has been suggested by Tsuji that this decrease is the effect of a dust cloud that is known to surround Betelgeuse. This cloud scatters the light from the surface of the star to give a halo, and it is the size of this halo that is measured by the speckle method. A familiar analogy is with the halo surrounding a street lamp on a foggy night. The scattering is less in the infrared, so the diameter apparently decreases with increasing wavelength, until at long wavelengths the true diameter of the star is measured. The Michelson interferometer gives a slightly larger diameter than the flux method, but this is probably within the error limits of the interferometer. The flux method provides further support for Tsuji's explanation because at still longer wavelengths, for example at 7 μm, it gives a large angular diameter. At these longer wavelengths there is an observed excess of infrared radiation above that expected from a star at the temperature of Betelgeuse. This excess radiation is present because the dust is heated to a temperature of over 1000 °C by stellar radiation at shorter wavelengths. The dust radiates at longer wavelengths because it is comparatively cool. For the star β Peg there is excellent agreement between the flux method and speckle interferometry, and this is especially pleasing because the flux method does not give an increased diameter at 7 μm, which suggests that this star is not surrounded by a dust cloud. From such comparisons it seems likely that the speckle method gives reasonably accurate diameters for stars showing very large discs. However, like the occultation method, it is limited to these stars, whereas the flux method is in principle applicable to stars of any disc size.

Even the Sun is not entirely free of a surrounding dust-cloud. It is only tenuous, but it can be seen during a total solar eclipse as the outer solar corona $(R/R_\odot > 5)$, and after sunset when the sun is more than about 18° below the horizon, as the zodiacal light. The zodiacal light can be seen only very rarely in this country but at a suitable mountain site in the tropics it can be a splendid and very impressive sight. Fig. 4.14 shows a photograph of it taken from an observing station near to La Paz in the Andes in Bolivia, at a height of 17 100 ft. This photograph was taken with a simple-lens camera using an exposure time of a few minutes, causing the stars to trail

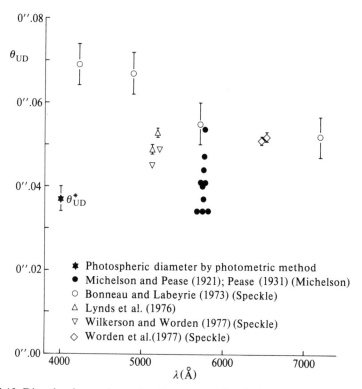

FIG. 4.13 Directly observed angular diameters of Betelgeuse plotted against wavelength. The filled circles are for Michelson interferometry, and all other symbols refer to speckle interferometry. The asterisk gives the angular diameter determined using the infrared flux method.

slightly. The atmosphere at this altitude is very transparent: the mountains on the horizon are at a distance of 160 miles.

The next step in this determination of the effective temperature of a star is to measure the integrated flux of radiation at the earth, $\mathscr{F}_E$. Fig. 4.15 shows the distribution of this radiation with wavelength for the star Sirius, which has a temperature of about 9800 °C. The very strong lines in this spectrum are due to absorption by hydrogen. The measurement of $\mathscr{F}_E$ requires the total area under this curve to be determined. It will be seen that in this spectrum there is energy down to a wavelength at least as small as 100 nm (1000 Å). A cool star such as Betelgeuse will not show any perceptible radiation at this wavelength, but a hot star might even have a maximum in its energy distribution here, with a strong flux at shorter wavelengths. For Sirius at least, a large proportion of the total radiation is in the visible region, say between 400 nm and 700 nm. This part of the spectrum can be measured relatively easily because the Earth's atmosphere is fairly transparent at these wavelengths. In contrast, the flux at longer wavelengths in

FIG. 4.14 Photograph of the zodiacal light obtained at a height of 17 000 ft in the Andes.

the infrared, particularly beyond 1 μm, can be measured only with great difficulty. There are two reasons for this. Firstly, radiation is absorbed at selective wavelengths by water-vapour in the earth's atmosphere. Infrared telescopes therefore have to be used at higher mountain sites, where the temperature is so low that almost all of the residual water-vapour has been frozen out of the atmosphere. Secondly, the flux of radiation is usually very small indeed, and extremely sensitive detectors cooled to low temperatures are needed. In addition, there is a further, somewhat unusual, difficulty. At the temperatures commonly found in observatories, all parts of the telescope are radiating strongly in the infrared, and the feeble radiation from the star has to be measured against an overwhelming contribution from the telescope. The situation can be likened to the rather fanciful one in which very faint stars are being measured in the visible region using a white-hot telescope. In addition to this, the sky is very bright at these wavelengths because the water-vapour and other constituents of the atmosphere are radiating very strongly. The result of this is that when an infrared telescope is pointed to a star, at 5 μm the signal from the star itself might be only 10^{-5} of the total signal, the remainder coming chiefly from the sky and the telescope. Nevertheless, the infrared region is a very important one in modern astrophysics, and in recognition of this the Science Research Council has funded a 3.8 m infrared telescope at Maunea Kea in Hawaii, at an altitude of 4200 m.

The near ultraviolet wavelength region 300 nm to 400 nm can be observed only with difficulty from the ground because of atmospheric absorption, much of which is due to dust and the scattering of light by the gaseous constituents. At wavelengths shorter than 300 nm, in the ultraviolet, absorption by atmospheric oxygen and ozone completely preclude obser-

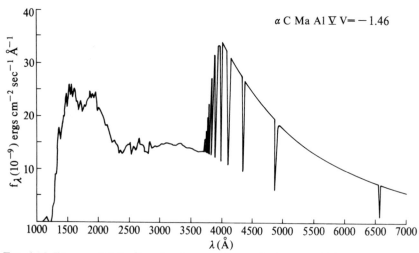

FIG. 4.15 Energy distribution in the spectrum of the star Sirius between the wavelengths of 100 nm (1000 A) and 700 nm (7000 A). (Courtesy of *Nature*.)

vation from the ground, and observations in this region must be made using either rockets or satellites. One of the first generation of such satellites was the ESRO TD-1 which carried an ultraviolet sky-survey telescope. This satellite was placed in circular orbit at an altitude of 550 km, and was arranged so that it could scan the whole sky in six months. During its life it observed a total of 31 215 stars over the wavelength range 135 nm to 255 nm. This satellite has been followed by another more advanced one called the International Ultraviolet Explorer (IUE) launched in January 1978 (and shown in Fig. 8.6). The telescope has a mirror made of beryllium, 45 cm in diameter, which directs light on to a high resolution spectrograph that records spectra over the wavelength region 115 nm to 320 nm. Fig. 4.16 shows an example of stellar spectra from this instrument, in this case of one of the hottest stars known, a white dwarf star HZ43, which has a temperature between 50 000 C and 100 000 C. Theoretical flux distributions for these two temperatures are given in the diagram.

The ultraviolet data from this satellite and its predecessor, in combination with ground-based data obtained with conventional telescopes and infrared data from Hawaii, are going to give most of the integrated stellar fluxes that are needed. These in turn, when combined with measured angular diameters, will give stellar temperatures. But in spite of the work of IUE there

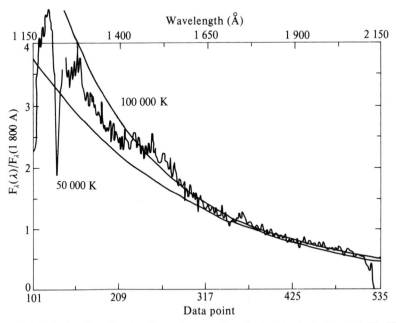

FIG. 4.16 Relative flux distribution with wavelength for the hot star HZ 43. The lower scale gives the data point number in the spectrum and the upper scale the wavelength in Ångstrom units. (Courtesy of *Nature*.)

will still remain a need for flux measurements at wavelengths shorter than 115 nm for very hot stars which are radiating significantly at these short wavelengths. These stars will have to wait for an EUV (Extreme Ultra-Violet) satellite.

One of the important stellar parameters mentioned earlier is the luminosity of a star, that is the total rate of radiation of energy by it. We already have, from the three sources already mentioned, the total rate of reception of energy from the star per square metre at the Earth. To convert this to the total rate of radiation from the star it must be multiplied by the area of a sphere which has the distance to the star as its radius. The accurate determination of this distance, which is related to the stellar parallax, is our next consideration.

Although we are really concerned with the distance to a star in kilometres, this distance is conventionally described by the parallax of the star, π, which is the angle subtended at the star by the radius of the Earth's orbit, measured in seconds of arc. This definition is used because this angle has to be measured in any direct determination of distance by triangulation, and of course it can always be converted to a distance knowing the radius of the Earth's orbit in kilometres. In principle, the measurement is very simple. Two photographs of the star and its surrounding field are obtained at an interval of six months, during which time the Earth will have moved half-way around its orbit. If the photographs have been taken at the appropriate times, the star will have moved relative to those near it in the sky (but actually far more distant along the line of sight). The distance moved is $2\pi \times F$, where F is the focal length of the telescope. A long-focal-length telescope is usually used for this measurement in order to make the displacement as large as possible. Even then the displacement to be measured is very small.

The nearest star is Proxima Cen, which has a parallax of 0.762 arcsec. If this star were to be measured with the Yerkes refracting telescope having a focal length of 19.4 m, it would show a total displacement of 0.15 mm. This could easily be measured, but in practice we need to work with stars that are at least 100 times as distant as this, with a accuracy of 1 per cent. To obtain improved accuracy, many photographs of a single star field are usually taken over a period of several years. As an example, in the measurement of the parallax of the star BD $+5°$ 1668 no few than 718 photographs have been taken by van de Kamp. However, the accuracy achieved for the vast majority of stars is still very poor. Sources of error are revealed by systematic differences between parallaxes determined for the same stars by different observatories. For example, there exists a long-standing difference between the parallax determinations by Allegheny and Yale Observatories of $\pi_Y - \pi_A - +0.0047$ arcsec. A consequence of this difference is that a star placed at a distance of 200 parsecs (652 light-years) by Allegheny Observatory will be placed at half this distance according to Yerkes Observatory.

III. The Orion region containing the Horsehead Nebula right of centre. The horsehead itself is an extension of the large dark cloud on the right which obscures more distant stars. The pink glow around the head is due to radiation from ionized hydrogen gas.

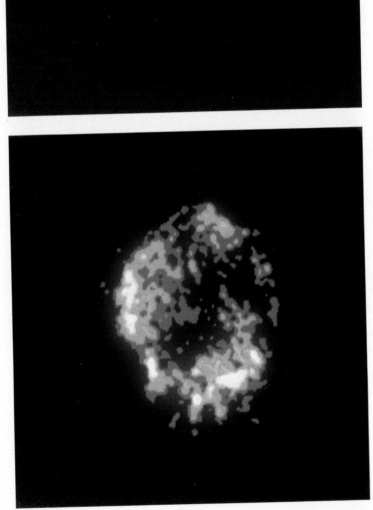

IV. X-ray image of the supernova remnant and bright radio source Cas A taken with the Einstein Observatory high resolution image (HRI). The extent of the emission region is believed to correspond to the present position of the supernova blast wave, with the more intense patches highlighting interstellar clouds or pre nova ejecta.

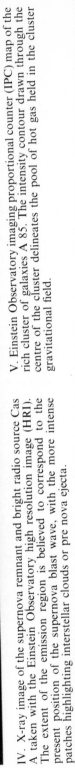

V. Einstein Observatory imaging proportional counter (IPC) map of the rich cluster of galaxies A 85. The intensity contour drawn through the centre of the cluster delineates the pool of hot gas held in the cluster gravitational field.

The distances of 11 500 stars have been determined by this direct parallax method, and the results catalogued by Jenkins. Some of the stars in this catalogue are even shown as having negative parallaxes. Of course, such parallaxes are physically impossible, but their presence demonstrates the existence of errors, which are always freely admitted by parallax observers. The distances of only ten stars are known with an accuracy of 2 per cent, all of them closer than 10 light-years. The distances of stars that are 100 light-years away from the Sun are known to only 20 per cent at best. This implies that the luminosity of a star at this distance is known to only 40 per cent accuracy.

An unsatisfactory feature of the present position is that, apart from the use of automatic plate measuring machines, fundamental techniques have scarcely advanced over the last 50 years. Now, there are before the astronomical community two proposals which, if implemented, could revolutionize the subject of stellar positions and distances. One of these proposals is for a satellite, called Hipparcos, which will measure positions and parallaxes with a far greater accuracy than is generally possible now. The satellite will achieve this by measuring very accurately the distance apart on the sky of pairs of stars that are separated by about 70 degrees. To do this, there will be two telescopes mounted on the satellite, pointing in directions 70 degrees apart, which direct the light from a star into an optical system that superimposes the two fields of view. At the focus of this system there is a grid made up of alternating opaque and transmitting strips, behind which is a photoelectric detector. The satellite is continuously rotating, causing the stars to move steadily across the grid. As they do so, the detector produces a varying output which is transmitted to a receiving station for recording and analysis. There are many advantages of this device over conventional ground-based telescopes, among them the absence of a disturbing atmosphere and the use of an efficient photoelectric detector instead of a photographic plate. This allows both increased accuracy and the possibility of observing a large number of stars over a short period of time. The expected life of the satellite will be about 2.5 years, providing in this time the parallaxes of between 50 000 and 100 000 stars down to a limiting magnitude of 13^m, to an accuracy of 0.002 arcsec. In addition, it will measure the proper motions of these stars to an accuracy of 0.002 arcsec yr^{-1}. During this short period it will have measured ten times as many stars as have been measured during last century, with a marked improvement in accuracy. Such a study will revolutionize our knowledge of stellar physics.

The purpose of the Hipparcos satellite is to measure a large number of stars with good accuracy. As an adjunct to this, Connes has considered what might be achieved by trying to improve ground-based parallax techniques. He concludes that if a specially designed photoelectric detector is used, and a special telescope is constructed, it should be possible to measure parallaxes to an accuracy of 0.0001 arcsec (0.1 milliarcsec), or even 0.04 milliarcsec if only 200 parallaxes per year are determined. This represents more than one order of magnitude improvement over Hipparcos, but of course a rate of working that is very much less rapid than that of the satellite. Such accurate data for selected stars would be of great importance, giving very accurate luminosities for nearby stars and extending the distance out to which good luminosities can be obtained, as well as giving accurate stellar diameters in kilometres. In addition, they would afford a check on the Hipparcos data which will otherwise remain quite unverified.

In making his proposal, Dr Connes had also in mind the possible detection of companions to stars which are so dark, perhaps because they are planets, that they are invisible in a telescope. The existence of such dark companions was first surmised by Bessel in 1844 when he announced that he had found small perturbations in the proper motions of Sirius and Procyon. The companions to these stars were seen much later through telescopes. A more modern counterpart of these examples is the star Ross 614, the binary character of which was deduced at McCormick Observatory in 1946 from a variable proper motion observed during a parallax programme. The hidden companion to this star, Ross 614B, was first seen and photographed by Baade in 1955 using the 200-inch telescope. The primary star of Ross 614 showed a total perturbation in right ascension of ± 0.25 arcsec, but much smaller amplitudes than this can be detected photographically. As an example, the star Ci 18 2354 shows a perturbation of amplitude of ± 0.002 arcsec in one coordinate. The perturbations are shown as a function of time over a period of 40 years in Fig. 4.17. An analysis of these data suggests a mass of $0.009 \, M_\odot$ for the companion, which must be very faint and still remains unseen in a telescope. The detection and investigation of these unseen companions is clearly an important and fascinating study, but one that requires observations made at intervals over long periods of time. Such

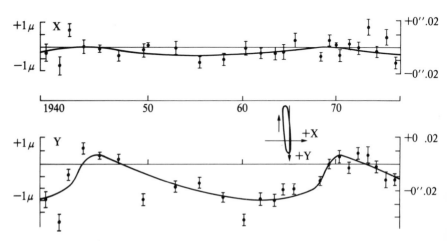

FIG. 4.17 Orbital displacements of the star Ci 18 2354 observed and calculated, over the interval 1939–76. (Courtesy of the *Astronomical Journal*.)

investigations are inappropriate for Hipparcos because of the shortness of its life.

We have described some of the new techniques of investigating stars in some detail in order to point the way to the new developments that will be taking place over the next decade. The future of stellar physics is again an exciting one, but it may be several decades before it is fully known how stars are born, live, and finally die. The Sun has always played an important part in stellar work because it is a test object that is so much easier to study than a remote faint star, and we conclude with a return to the Sun and a brief reference to an new aspect of solar physics.

It is usually assumed that the Sun is a stable star, but there is now a variety of evidence to show that it actually undergoes very small oscillations in size and brightness. One way of investigating this kind of change is to measure very accurately the wavelength difference between an absorption line in the solar spectrum and a nearby absorption line formed in the Earth's atmosphere. As a consequence of the Doppler effect, this wavelength difference is sensitive to the relative speed of the observer and the Sun. Fig. 4.18 shows a plot of the observed difference between the solar line FeI 629.78 nm and a neighbouring terrestrial line made over a period of a few hours. In this plot, the slowly changing wavelength difference is due to the changing component during the day of the observer's speed towards the Sun caused by the Earth's rotation. The fine structure on this diagram demonstrates that the Sun is oscillating, or breathing gently, with a period of about 5 minutes. Extremely accurate measurements are needed. Here, the

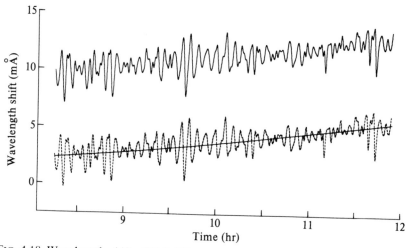

FIG. 4.18 Wavelength shift of FeI 448.98 in the solar spectrum. The upper curve shows the observed wavelength relative to a laboratory source. The lower curve shows this displacement superimposed on the contribution due to observer motion, the motion of the Earth being calculated from Newcomb's theory.

accuracy of measurement is such that it would be possible to detect fairly easily the motion of a walking person through the Doppler shift in the wavelength of light reflected from him. Oscillations may also be seen directly through variations in the size of the Sun, but the amplitude of these oscillations are not compatible with those expected from measurements of velocity changes. There is no doubt that here is yet another new technique for investigating the structure of the Sun. It is possible that these solar oscillations will allow the internal structure to be probed directly. Just as seismic waves in the Earth have allowed geophysicists to determine the mantle/core structure of the Earth, solar oscillations may provide clues to the inner structure of the Sun.

Notes and sources

Within the last decade the Science Research Council has provided several new ground-based telescopes of exceptional size and quality. These include the UK 48-inch Schmidt telescope in Australia (which is actually a *camera*, not a telescope), a half-share in the 150-inch Anglo-Australian telescope, and the recently completed 3.8 metre infrared flux collector at Maunea Kea in Hawaii. These instruments are described in the following publications, Cannon, R. D., *Royal Greenwich Observatory Bulletin*, No. 182, p. 283 (1976); Wampler, E. J. and Morton, D. C., *Vistas Astron.* **21**, 191 (1977); Humphries, C. M., *Sky and Telescope* **56**, 22 (1978). The ESRO TD-1 ultraviolet sky survey telescope and the International Ultraviolet Explorer satellite are described by Boksenberg *et al.*, *Mon. Not. R. astr. Soc.* **163**, 291 (1973) and Boggess *et al.*, *Nature, Lond,* **275**, 372 (1978).

The dark globules were first discussed by Barnard, *Astrophys. J.* **49**, 1 (1919). They were not studied again for nearly sixty years until the work of Bok, *Publ. Astron. Soc. Pacific* **89**, 597 (1977). Bok has remarked that because of their extreme darkness they are striking objects when viewed in a large telescope.

A description of the solar neutrino experiment may be found in many publications, but the most comprehensive account of the details of it, with the implications of the experimental results, may be found in the book by Eddy, *The new solar physics*, AAAS Selected Symposium 17 (1978) Westview Press, Colorado, USA. This book also contains an account of the present state of work on solar oscillations.

An account of the basic principles of stellar evolution, dated 1970, is given by R. J. Tayler in his book *The stars: their structure and evolution*, Wykeham Publications (London) Ltd. The original work of Labeyrie on speckle interferometry is described in his paper in *Astron. Astrophys.* **6**, 85 (1970); Labeyrie has also given a review of interferometric methods in astronomy in *Ann. Rev. Astron. Astrophys.* **16**, 77 (1978). Much of the modern development of the occultation method for determining stellar angular diameters has been done by D. S. Evans and R. E. Nather (e.g. Nather, *Astron. J.* **78**, 583 (1973); Nather and Evans, *Astron. J.* **78**, 575 (1973)). The history of the Michelson stellar interferometer is discussed by DeVorkin (*J. Hist. Astron.* **6**, 1 (1975)), and the history of the development of the stellar intensity interferometer is described by Hanbury Brown in his book, *The intensity interferometer*, Taylor and Francis, London. The infrared flux method for determining stellar temperatures and angular diameters is described by Blackwell, Shallis, and Selby in *Mon. Not. R. astr. Soc.* **188**, 847 (1979). Pierre Connes discusses the future of ground-based astrometry in his paper, 'Should we go to space for parallaxes?' (*Astron. Astrophys.* **71**, L1 (1979)). In this he states, 'there is not one shred of evidence anywhere to prove narrow field astrometry is bumping up against a "natural" atmospheric limitation'.

5

The X-ray Universe

K. A. POUNDS

Introduction

Our only direct information on the nature of the stars and galaxies throughout the Universe comes in the form of electromagnetic radiation. For thousands of years this input was limited to the visible region, the narrow waveband of the electromagnetic spectrum lying between 0.35 and 1 micrometre, to which our eyes are sensitive and the Earth's atmosphere is transparent. The development of radar techniques during the Second World War led to the exploitation of the other, long wavelength, atmospheric 'window', and gave birth to radioastronomy, a field of research bearing increasingly rich fruits from the early 1950s. Then came the 'Space Era', heralded by *Sputnik 1* in October 1957, which finally opened up the entire electromagnetic spectrum to astronomers. Fig. 5.1 illustrates this point and indicates how rocket- and satellite-borne equipment can now be used to observe the Universe in the previously hidden wavebands of γ-rays, X-rays, ultraviolet light, and the far infrared.

Of all these, the X-ray waveband has so far been the most productive, contrary to earlier expectations. From the vantage-point of the late 1950s ultraviolet studies of hot stars and of the interstellar medium, and the detection of penetrating γ-rays from cosmic-ray interactions throughout the Milky Way were expected to be the most valuable fruits of space observations. Although X-rays had been detected from the Sun's corona by Herbert Friedman and colleagues at the US Navy Research Laboratory in the earliest space astronomy experiments using captured V2 rockets, theorists predicted little future for X-ray astronomy beyond the detection, well into the future, of the much fainter signals from other, relatively nearby stars. How differently events turned out, I will attempt to describe in the present chapter.

The story of cosmic (or non-solar) X-ray astronomy begins in 1962, with the chance discovery of a bright source in the Scorpius constellation during

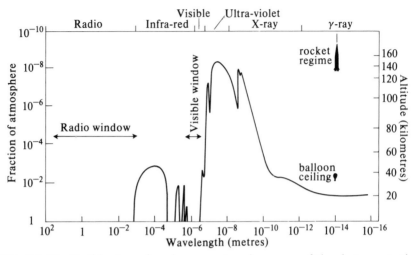

FIG. 5.1 The Earth's atmosphere is opaque to a large part of the electromagnetic spectrum. Radiation from outer space penetrates approximately to the depth shown, reaching the ground only in the visible and radio 'windows' and in the near infrared.

a brief rocket flight primarily intended to search for fluorescent X-rays from the Moon. The discovery of a remarkably powerful source in Scorpius was announced to an initially sceptical world by Riccardo Giacconi and his colleagues working at American Science and Engineering, an offshoot of the MIT Science Faculty. All doubts evaporated during the following year, with new sightings of Scorpius X-1 by the ASE and NRL groups, and of a second bright source, in Taurus. The latter, Tau X-1, was rapidly identified with the Crab Nebula supernova remnant in a famous NRL rocket experiment in 1964, during which the lunar occultation of Tau X-1 was observed which both located its accurate position and showed the X-ray emission to extend across the visible nebula (Fig. 5.2).

Progress through the rest of the 1960s was maintained, with further rocket and balloon flights extending the list of detected sources to between 25 and 30, including a number discovered in the southern sky by rocket surveys from Woomera, South Australia conducted by the University of Leicester group. Fig. 5.3 shows a typical rocket payload of this period. Although in 1966 the prototype Sco X-1 source was successfully identified with a faint (~13th magnitude) blue star, most X-ray sources remained unidentified and X-ray astronomy was still well removed from the mainstream of astrophysics research. However, the launch, on 11 December 1970, of the first X-ray astronomy satellite was to lead the subject into an entirely new and richer phase.

The new satellite was called *Uhuru* ('freedom' in Swahili) in recognition of its launch on Independence Day in Kenya, from whose eastern coast it was injected into a 500 km equatorial orbit. It carried a simple payload, basically similar to those flown in the rocket surveys of the late 1960s: two 840 cm^2 proportional gas counters looking out in opposite directions in the spin plane of Uhuru, each sensitive in the photon range 2–20 keV and with fields of view of $0.5° \times 5°$ and $5° \times 5°$, respectively. The main difference was the long exposure afforded by *Uhuru*, compared with the 5 to 10 minutes of a typical rocket flight, which gave a much improved sensitivity, as well as a more complete sky coverage and (very important, as it turned out) the ability to monitor X-ray variability.

Following *Uhuru*, the next satellite devoted to X-ray astronomy was *Ariel 5*. This UK satellite was launched on 15 October 1974, again from Kenya, and carried a payload of 6 experiments (see Fig. 5.4), including a new sky survey instrument from Leicester. Continuous operation of the *Ariel 5* survey up to the present time has further extended our X-ray map of the sky, with the current distribution of some 300 sources shown in Fig. 5.5. This

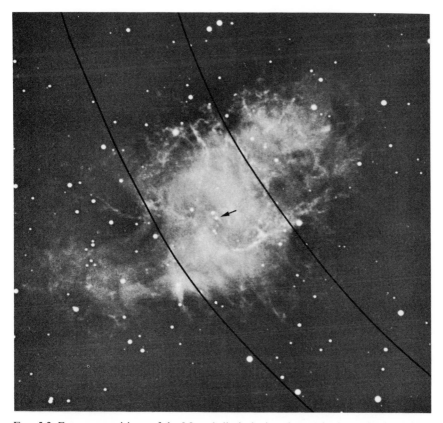

FIG. 5.2 Extreme positions of the Moon's limb during the gradual occultation of the X-ray source Tau X-1 in 1964. This observation clearly identified the X-ray source with the Crab Nebula supernova remnant, the centre of the X-ray emission lying close to the pulsar NP 0532.

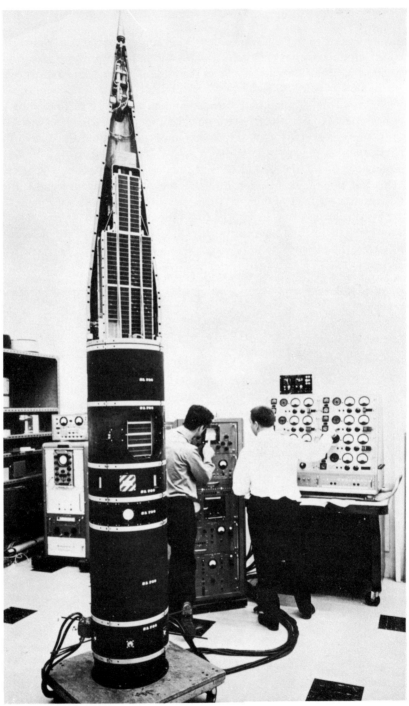

FIG. 5.3 *Skylark* rocket payload used to search for X-ray sources in the southern hemisphere in the period 1966–71. One of two back-to-back large area X-ray proportional counters is shown mounted in the rocket nose-cone section.

FIG. 5.4 The 130 kg *Ariel 5* satellite undergoing final preparations before launch in October 1974. The outer window of the Leicester Sky Survey Instrument is seen at the top left.

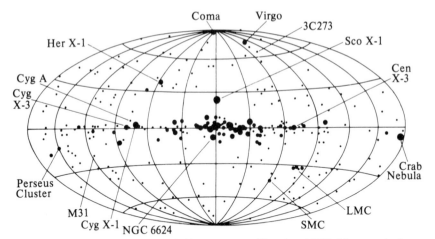

FIG. 5.5 *Ariel 5* map of 297 cosmic X-ray sources (October 1978). The map is shown as an equal area projection in galactic co-ordinates with the Milky Way along the equator, galactic centre in the middle, and galactic north pole at the top. The (log of the) X-ray intensity of a source is approximately represented by the diameter of the corresponding filled circle.

map is in galactic co-ordinates. It shows a concentration of X-ray sources along the galactic plane, especially within the 'galactic bulge', but also a large number spread uniformly across the whole sky. Further study has shown many of the latter to be extragalactic sources, most lying well beyond the limits of our local galaxy. In the following sections the nature of both the galactic and extragalactic sources will be described.

X-ray binary stars

The ability to monitor X-ray variability over long periods brought the most outstanding achievement of the *Uhuru* satellite: the discovery of X-ray binary stars. The discovery came quickly, first with the detection of 'pulsed' X-rays from one of the bright sources in the southern Milky Way, Centaurus X-3, followed by the further discovery that the X-ray source 'turned off' every two days or so, and then that the 4.8 s pulse period varied in phase with the 2.09 day 'eclipse'. That it was indeed an eclipse of the X-ray source by a binary stellar companion was confirmed by the sinusoidal nature of the in-phase pulse period variation, this being attributable to the Doppler effect of the X-ray star moving with high velocity around its companion star (Fig. 5.6).

Uhuru observations over the following year led to a catalogue of 161 X-ray sources, some two-thirds of which lay close to the Milky Way and were evidently of galactic origin. Moreover, the latter were typically variable on time-scales of days or less, suggesting the majority to be stellar systems rather than supernova remnants. Regular X-ray variability, coupled with an increasing number of successful optical identifications, showed that many of the galactic sources were probably binary systems rather like Cen X-3. Table 5.1 summarizes the main properties of the identified galactic sources in the third *Uhuru* (3U) catalogue. An outstanding feature of *all* the sources listed in Table 5.1 is their extremely high luminosity. Typically these range from 10^{29} to 10^{31} watts compared with a *total* radiated power of the Sun of $\sim 10^{27}$ watts. Clearly these X-ray stars were unusual objects indeed.

A model of the X-ray binaries which is now widely accepted was soon proposed. Fig. 5.7 illustrates the situation. A pair of stars lie sufficiently close for mass transfer to occur and—crucially—the accreting star, which is also the X-ray star, is extremely small. The latter property ensures ample gravitational potential to power the X-ray binary. Fig. 5.7(a) shows a low-mass X-ray binary, where both stars are of order of the Sun's mass. Here mass transfer occurs as a result of the companion star filling its own half of the first common equipotential surface. Gas then flows through the

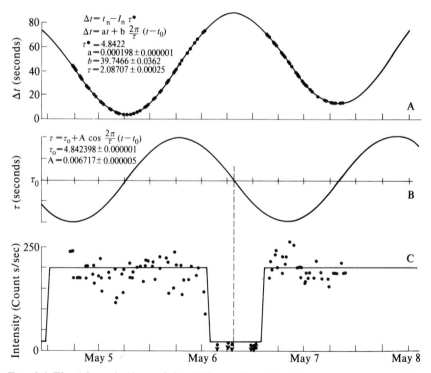

FIG. 5.6 The 4.8 s pulsations of Cen X-3 were found by *Uhuru* to vary sinusoidally and in-phase with the binary period. The graph shows (A) the measured (relative) time of arrival of the pulse train, (B) the instantaneous pulse period, and (C) the overall intensity, during an extended observation in May 1971.

TABLE 5.1

Galactic X-ray sources identified in third Uhuru (3U) catalogue

	3U source	Distance (Kpc)	X-ray power at 2–10 keV (watts)	Remarks
A. Supernova remnants				
Tycho's Nova	0022+63	~5	~5×10^{28}	SN of 1572
Crab Nebula	0531+21	~2	~10^{30}	including pulsar NP 0532
Puppis A	0821–42	~2	~10^{28}	10^4–2×10^4 years old associ-
Vela X	0833–45	~0.5	~10^{27}	ated with pulsar SN of
Cassiopeia A	2321+58	~3.4	~10^{29}	1653–71
B. Close binaries				
Vela XR-1	0900–40	~1.3	~10^{29}	9.0 day eclipsing binary
Centaurus X-3	1118–60	~5	~10^{30}	2.1 day eclipsing binary
Hercules X-1	1653+35	2–6	10^{29}–10^{30}	1.7 day eclipsing binary
—	1700–37	~1.7	~10^{29}	3.4 day eclipsing binary
Cygnus X-1	1956+35	~2	~5×10^{29}	5.6 day optical period
C. X-ray stars				
X Perseus	0352+30	~0.4	~5×10^{26}	Probable optical period 580 days
Scorpius X-1	1617–15	~0.5	~5×10^{29}	Optical counterpart V 818 Sco
Cygnus X-2	2142+38	1–2	1–5×10^{29}	14th magnitude F2 star
Cygnus X-3	2030+40	~10	~5×10^{30}	4.8 hour X-ray period, infrared and radio source
D. Globular clusters				
M 92	1736+43	—	—	Not confirmed in 4U catalogue
NGC 6441	1746–37	~9	~10^{30}	
NGC 6624	1820–30	~6	~10^{30}	
M 15	2131+11	~10	~10^{29}	

connecting neck into the gravitational domain of the compact star. What happens thereafter is primarily determined by the need to get rid of the angular momentum of this material. A part is probably ejected from the system while some returns to the original star, but a substantial fraction sinks on to the compact star via an accretion disc. This fraction acquires a large amount of kinetic energy from the strong gravitational field of the compact star, resulting in very high temperatures in the inner part of the accretion disc and emission of the observed powerful flux of X-rays. Both the X-ray luminosity, L_X, and the temperature of the source, T, are directly dependent on the degree of compactness of the X-ray star. For a rate of accretion of $\dot{m}$ on to a star of mass M_X and radius R_X,

$$L_X \sim \dot{m}v^2 \sim \dot{m}G \left(\frac{M_X}{R_X} \right) \tag{5.1}$$

$$\sim 10^{26} \left(\frac{M_X}{M_\odot} \right) \left(\frac{R_\odot}{R_X} \right) \left(\frac{\dot{m}}{10^{-8} \, M_\odot \, \mathrm{yr}^{-1}} \right) \mathrm{watts}$$

and

$$T \sim \frac{\alpha M_p}{k}\left(\frac{M_X}{R_X}\right) \tag{5.2}$$

$$\sim 10^7 \alpha \left(\frac{M_X}{M_\odot}\right)\left(\frac{R_\odot}{R_X}\right) \text{Kelvin}$$

Here α is an efficiency factor for the heating process, having a value $\sim 10^{-5}$ in the viscous heating likely to dominate in an accretion disc. Thus, to obtain the high temperatures typical of X-ray binaries ($T \gtrsim 10^7$ K) and an X-ray luminosity as high as 10^{31} watts, with the maximum mass transfer that will not 'swamp' the source ($\dot{m} \sim 10^{-8}$ $M_\odot$ per year), we see that

$$\left(\frac{M_X}{R_X}\right) \gtrsim 10^5 \left(\frac{M_\odot}{R_\odot}\right).$$

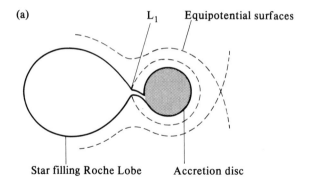

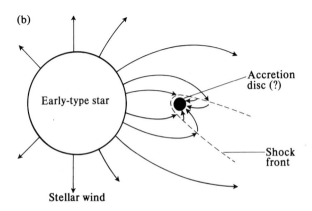

FIG. 5.7 Mass transfer in close binary X-ray source by (a) Roche lobe overflow, and (b) accretion of stellar wind emanating from a luminous early-type star.

Only one kind of star is so compact, viz. a neutron star, while a black hole of stellar mass could also provide the required gravitational 'potential well'. The existence of neutron stars, first predicted in 1934 shortly after Chadwick discovered the neutron, had, of course, been established with the discovery of radio pulsars in 1967, but the interaction of such exotic objects with a binary companion now offered X-ray astronomers a powerful new way of studying the properties of matter at nuclear densities. Already such studies have provided unique information on the structure, mass distribution, and evolution of neutron stars, while a recent balloon observation by Joachim Trümper and his colleagues at the Max Planck Institute in Munich has yielded a direct measurement of the intense magnetic field ($\sim 10^8$ tesla) associated with the neutron star in the X-ray binary Hercules X-1.

Hercules X-1 is probably well described by the model of Fig. 5.7(a), its binary companion being an F5 star of about one solar mass. In the case of Cen X-3, the optical identification with an O 6.5 supergiant star by Krzeminsky suggests the mass transfer is more likely to be via an intense stellar wind from the optical primary, as depicted in Fig. 5.7(b). In both cases, however, the X-ray power is derived from the intense gravitational field of the neutron star, with the rapid X-ray pulsations (4.8 s period for Cen X-3 and 1.25 s period for Her X-1) produced by the channelling of the accreted material on to the magnetic poles of the rotating neutron star by its intense magnetic field (Fig. 5.8).

Any discussion of the subject of X-ray binaries naturally leads to the intriguing question of experimental evidence for the existence of black holes. It seems clear that the best chance of such verification will come from the interaction of a black hole with a nearby stellar companion in a binary system. This would probably appear as a powerful X-ray source, perhaps like Cen X-3 but without the regular pulsations. Cygnus X-1 is such a source, identified with a supergiant O 9.7 star with an apparent binary period of 5.6 days. The implied mass of the compact X-ray star is very high, between 5 and 15 M_{Sun}, and this lies well above the theoretical upper limit for a neutron star of $\sim 3.5\ M_{Sun}$. The great majority of astrophysicists find this compelling evidence for the existence of a stellar mass black hole in Cygnus X-1, although the entrenched disbeliever still has one or two possible avenues of escape, for example in the form of a triple star system. But for this there is *no* evidence in Cygnus X-1. More examples of X-ray sources with the properties of Cyg X-1 are clearly needed to remove entirely any contrived alternatives to the black hole interpretation. The new UK-6 satellite, launched in May 1979, may provide the final proof of stellar black holes, equipped as it is to study X-ray sources in much greater spectral and temporal detail than has been possible hitherto.

Two final comments should be made before leaving the X-ray binaries. The first is to report the recent discovery, by the *Ariel 5* satellite, of 2–10 keV X-rays from an X-ray binary containing neither a neutron star nor a black hole. The dwarf nova, SS Cygni, was observed during an optical

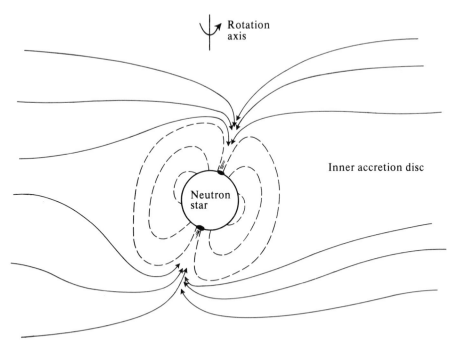

FIG. 5.8 Accreting gas channelled on to poles of magnetic neutron star, forming intense X-ray 'hot-spots'. Rotation about an axis oblique to the magnetic axis is believed to explain the rapid X-ray pulsations of sources such as Cen X-3 and Her X-1.

outburst, when the X-ray flux was seen to vary inversely with the optical brightness (Fig. 5.9). Dwarf novae are known to be close binaries of low mass, with a white dwarf secondary. The observation of the (relatively weak) X-radiation ($L_X \sim 10^{26}$ watts) has been attributed to the shock heating of accreted gas falling directly on to the white dwarf, the relevant value of α in equation (5.2) being ~ 0.1, thereby yielding the observed $T \sim 10^7$ K for the lower value of M/R appropriate to a white dwarf,

$$\sim 10^3 \left(\frac{M_\odot}{R_\odot} \right).$$

Secondly, there are the bright X-ray transients, a phenomenon which created much interest during the early days of *Ariel 5* in 1974/5. Bright and apparently short-lived sources had been seen by *Uhuru*, but the data were generally limited and no optical identifications were made. In fact, the very first transient, named Cen X-2, was discovered during the earliest *Skylark* survey flights from Woomera in April 1967, but, by the time the transient nature of the source had been realized months later, it was too faint to begin an effective optical search. The nature of the *Ariel 5* project, in which data were available to the experimenters within a few hours (see Fig.

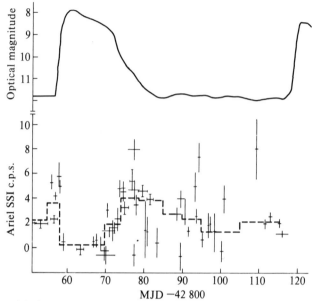

FIG. 5.9 *Ariel 5* observations of the dwarf nova SS Cygni confirming its identification as an X-ray source of moderate power and demonstrating an anti-correlation between the X-ray and optical emission during the optical outbursts.

5.10), overcame this difficulty and led, quite rapidly, to the first identifications of X-ray transients.

Fig. 5.11 shows a 'sky survey' scan along the galactic plane during the first weeks of *Ariel 5* operation. The intense sources in the general direction of the galactic centre (the 'bulge' sources) stand out, together with other galactic sources also previously listed by *Uhuru*. Little interest was aroused by the faint source TrA X-1 until it was noticed to be brightening daily. A careful check of its position showed that it was absent from the *Uhuru* catalogue. Although the Sun lay close to the region of TrA X-1 in November, a telex from Leicester to Paul Murdin of the RGO, then assisting in the commissioning phase of the new Anglo-Australian 3.9-metre telescope at Siding Springs, produced the photograph shown in Fig. 5.12(a). This early, contemporary photograph proved crucial to the successful optical identification of TrA X-1, since over the following months, as the X-ray intensity peaked and then fell away, further optical photographs revealed just one of the many stars in the X-ray error box to fade in concert. Fig. 5.12(b) shows a deep exposure in June 1975, with the optical counterpart of TrA X-1 now two magnitudes fainter than its initial magnitude six months previously. As with the 'steady' X-ray binaries, the additional optical information on Tra X-1 and several similar X-ray transients has shown these also to be binary systems. TrA X-1, for example, appears to be a low-mass binary, not unlike Her X-1, but the primary star probably does not quite fill its Roche lobe. When, owing to temporary expansion of the primary, the Roche lobe is filled, mass transfer to the compact companion is suddenly 'switched on' and the X-ray source appears. Later, with relaxation of the primary, the gas supply is cut off again and the X-ray source—and that part of the optical flux arising from X-ray heating of the primary and from the accretion disc—fades away as the disc decays.

The comparison of this scenario with that now thought to apply to most optical classical novae was quickly clarified with the occurrence of 1975 Nova Cygni. Though the brightest optical nova recorded for approximately 40 years, no X-rays were discernable from 1975 Nova Cygni, implying an X-ray to optical power ratio less than 10^{-7} of that of the brightest 'X-ray nova', 1975 Nova Mon (also discovered by *Ariel 5*). The explanation for this remarkable difference may be sought in the nature of the compact star in the two systems (see eqn. 5.2), this being, presumably, a white dwarf in Nova Cygni, but a much more compact neutron star (or, possibly, black hole) in the bright X-ray transients.

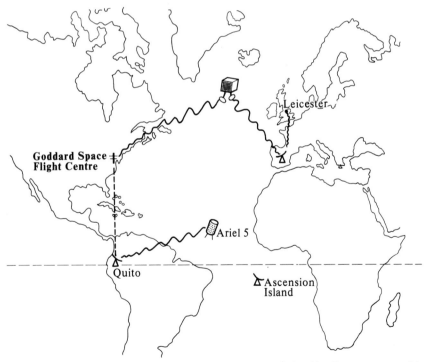

FIG. 5.10 *Ariel 5* Command and Data link. The equatorial orbit allows contact with stations at Quito (or Ascension Island) each ∼97 minutes, when stored data is relayed to ground and spacecraft and experiment commands transmitted to *Ariel 5*. A cable-microwave-satellite link joins the NASA ground stations to the Operations Centre at the Appleton Laboratory. Sky Survey data is available for inspection at the Leicester computer terminal within a few hours of its recording in space.

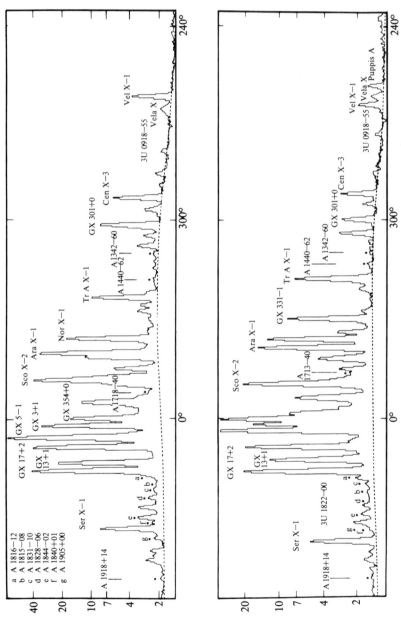

FIG. 5.11 Galactic plane seen by the *Ariel 5* Sky Survey Instrument in November 1974. To the right of the intense galactic centre sources is seen TrA X-1 shortly after its first appearance. The upper and lower plots are from the two separate sections of the SSI detector array.

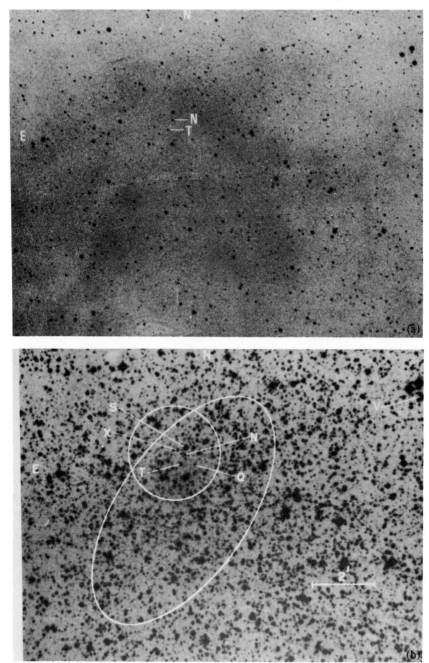

FIG. 5.12 (a) Photograph of the field of TrA X-1 with the Anglo-Australian 3.9-metre telescope on 15 December 1974. Star *N* is comparable to star *T* at ~17th magnitude. (b) The same field photographed in a deeper exposure with the UK Schmidt telescope on 30 June 1975. Star *N* is now comparable to star *Q* at ~19th magnitude.

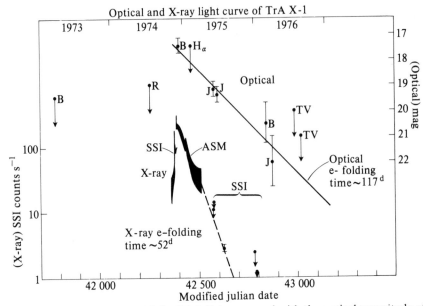

FIG. 5.13 The total rise and fall of TrA X-1 compared with the optical magnitude of star *N*, confirming its identification with the X-ray transient source.

X-rays from supernova remnants

As noted earlier, the Crab Nebula (remnant of the supernova of AD 1054) was the first cosmic X-ray source to be optically identified, in 1964. Many subsequent observations have been made which show that the Crab emits X-rays over a wide energy band (~0.5 keV to beyond 1 MeV), from a region coincident with the source of visible light, and polarized to a level of ~15 per cent. Further, the X-ray emission region surrounds the Crab pulsar, NP 0532, and it seems clear that a continuous supply of highly relativistic electrons accelerated by the pulsar is responsible for the visible and X-ray emission by the same, non-thermal process, namely synchrotron radiation. In this process relativistic electrons (of energy *E*, eV) interact with the *in situ* magnetic field (*H*, tesla) to produce a broad spectrum of photons of frequency near *v*, where

$$v \simeq 0.1 \, HE^2 \text{ Hz} \tag{5.3}$$

For the Crab Nebula, *H* has been measured to be ~3 × 10⁻⁸ tesla, requiring electrons of $E \sim 10^{13}$–$10^{14.5}$ eV to yield X-rays in the observed range 1–100 keV. The synchrotron process is so efficient for such extreme electron energies that the electron lifetimes are very short. Roughly, their effective lifetime is *τ*, where

$$\tau \simeq 5 \times 10^5 \, E^{-1} H^{-2} \text{ s} \tag{5.4}$$

In the Crab Nebula, τ ranges from ~ 2 years to ~ 1 month for $E \sim 10^{13}$–$10^{14.5}$ eV and the existence of an active pulsar is essential for continuing X-radiation from the Crab. Conversely, it is interesting to note that a substantial part of the rotational energy loss of NP 0532 (with a period decreasing from ~ 33 ms at some 15 μs per year) is being used to power the X-ray source.

Despite its remarkable nature, the Crab Nebula is apparently a very untypical supernova remnant. Only the pulsar PSR 0843-30 in Vela (with the second shortest known period of 84 ms) seems to have a similar, but much weaker, extended synchrotron X-ray source. The Cygnus Loop is much more typical of the dozen or so other X-ray emitting supernova remnants. Fig. 5.14 shows a recent X-ray picture of the Cygnus Loop obtained with a rocket-borne flight of an imaging X-ray telescope. This experiment, carried out in July 1977 by the Leicester and MIT groups, was notable in yielding the first cosmic X-ray image obtained with the powerful Wolter I telescope technique, now being employed on the Einstein Observatory (of which more later). The Cygnus Loop is older than the Crab, about 2×10^4 years old, and its X-ray image has an evident shell-like structure. Also, its X-ray spectrum appears with the characteristic shape of a thermal source, with a temperature of $\sim 4 \times 10^6$ K. The interpretation placed on these data is that the X-rays are due to heated interstellar gas, energized by the advancing blast-wave from the initial supernova explosion. On this basis the outer limit of the X-ray image lies close to the present position of this blastwave, while the patchy appearance in the image probably highlights the cloudy nature of the initial interstellar medium. The absence of a clear X-ray shell in the southern part of Cygnus Loop (furthest from the galactic plane) suggests a lower interstellar density there, with relatively little impediment to the blast-wave and, correspondingly, minimal heating and X-ray emission. X-ray images of this kind are clearly going to be of great value in the study of supernova remnants in the next few years, providing information on the evolution and energetics of each remnant as well as the interstellar medium into which they are propagating.

Clusters of galaxies

A second outstanding and equally unexpected discovery by *Uhuru* was of powerful X-ray emission from rich clusters of galaxies. Even with the modest spatial resolution of the *Uhuru* survey detectors ($\sim 0.5°$) it was evident that the emission from the nearer rich galaxy clusters in Virgo, Perseus, Coma, and Centaurus was extended over a degree or so, corresponding to a linear size of $\sim 10^6$ light years. It is remarkable that powerful X-ray emission had thus been found in association with both the smallest and largest entities known to exist beyond the solar system. Subsequent observations, particularly by *Ariel 5*, have shown X-ray emission to be a property of *all* clusters, the amount of X-radiation being roughly dependent on the number of galaxies and their degree of concentration in the cluster (Fig. 5.15). The origin of the X-radiation was at first unclear, although it was evidently too intense—by one to three orders of magnitude—to be simply the sum of the emission from the individual galaxies. An observation by the Mullard Space Science Laboratory group, with *Ariel 5*, provided a breakthrough. Fig. 5.16 shows their result for the X-ray spectrum of the Perseus cluster, revealing a clear feature near 7 keV. The interpretation of this feature, similar to that

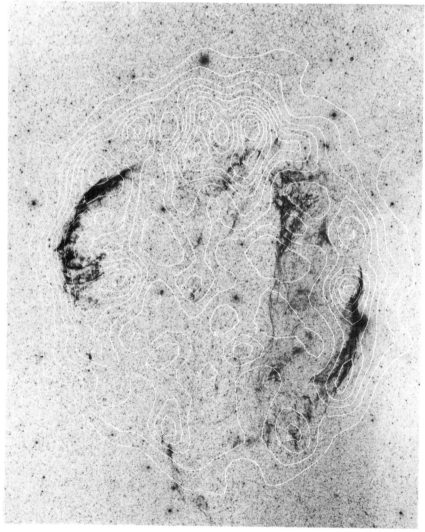

FIG. 5.14 X-ray image of the Cygnus Loop supernova remnant superimposed on a Palomar Sky Survey print. The shell structure of the 0.5–1.5 keV image is clearly seen.

often seen in solar flares, as characteristic radiation of highly ionized iron, was strong evidence for the thermal nature of the X-radiation. The temperature of $\sim 10^8$ K implied by the overall spectral shape is sufficient to fully ionize all elements lighter than iron and leave only the K shell electrons of that element intact. Although the *Uhuru* and *Ariel 5* observations did not have the angular resolution to demonstrate the fact, it seemed likely that the source of the X-radiation was a tenuous hot gas filling the vast space between the galaxies. Beautiful confirmation of this view has very recently been obtained with the Einstein Observatory X-ray telescope (Plate 5).

The origin of the intracluster gas and the manner of its heating are at present unclear, although it seems probable that at least a part is due to 'primordial gas' falling into the gravitational potential well of the cluster and thereby gaining the necessary thermal energy. As no direct observational evidence exists for this primordial *inter*cluster gas the future X-ray observations of clusters will be especially interesting. A second possibility, that a substantial fraction of the X-ray-emitting gas has been ejected from the individual galaxies, is perhaps supported by the detection of the iron feature (now seen in five of the brightest X-ray clusters) since it would be difficult to account for the necessary relative abundance of iron (about one half of that in the solar system) in the primordial gas, since this is believed to be the pure remnant of the initial Big Bang. X-ray observations over the next few years should solve this point, together with the very important question as to how clusters—and their constituent galaxies—evolve with time. Present data still leaves these questions open.

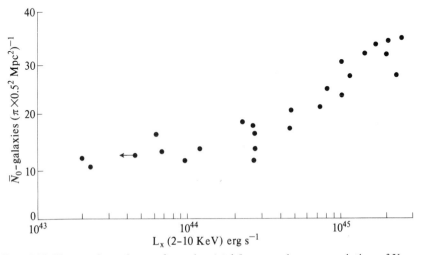

Fig. 5.15 X-ray galaxy clusters from the *Ariel 5* survey show a correlation of X-ray luminosity with the central galaxy density of the cluster, in turn determining the gravitational 'well' which holds the hot X-ray-emitting gas.

Active galaxies

As noted above, clusters of galaxies were established as an important class of
extragalactic X-ray source by *Uhuru*. The 3U catalogue included 11 such
identifications. Following the launch of *Ariel 5* in October 1974, the
extended sky survey began to add to this list and the number of X-ray clusters
had grown to 50 by 1978, representing over a third of all the X-ray sources
at high galactic latitude. The *Ariel 5* survey, furthermore, led to a second
major class of extragalactic source being established, following a pro-

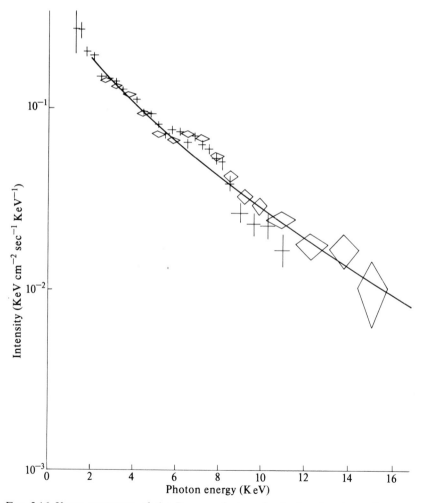

FIG. 5.16 X-ray spectrum of the Perseus cluster of galaxies showing a feature at
~ 7 keV attributed to thermal emission from highly ionized iron at $\sim 10^8$ K.

gramme of joint X-ray and optical studies by the Leicester *Ariel 5* team and astronomers at the RGO and the Anglo-Australian Observatory. This second class covers the X-ray active galaxies, including QSOs (or quasars), Seyferts, BL Lacertids, and emission line galaxies, which together now form a group of comparable number and X-ray power to the clusters.

The study of X-ray galaxies has been an interesting demonstration of the rapid convergence of X-ray astronomy and the mainstream of astrophysics. In the 3U catalogue only 4 active galaxies are identified, viz. the brightest QSO (3C 273), the nearby radio galaxy (Cen A), the optically bright Seyfert (NGC 4151), and, albeit with a very uncertain X-ray location, an irregular galaxy (M 82). Faced with some 50 new unidentified X-ray sources at high galactic latitude, the Leicester–RGO–AAO team embarked on optical studies of the X-ray error boxes. An early result was the identification of the listed Seyfert galaxy NGC 3783 with 2A 1135-373. Following this, optical spectra of the brightest galaxy or galaxies in each error-box produced an unexpected crop of (previously unknown) Seyferts. Fig. 5.17 shows an early example, with the galaxy ESO 113-IG 45 indicated in the western half of the error-box of 2A 0120-591. A range of optical exposures of this galaxy in Fig. 5.18 shows very clearly the dominantly bright nucleus which is typical of active galaxies. In fact, having determined its red-shift (and distance) the optical power of this previously unknown Seyfert was found to be the greatest in the southern hemisphere and comparable with a low luminosity QSO. A second new Seyfert, established from the search of *Ariel 5* error-boxes, MCG-2-58-22, is a further notable X-ray–optical discovery, having hydrogen emission lines broader than any other known Seyfert apart from 3C 382.

As the number of X-ray Seyferts increased, it became possible to look for some pattern within the X-ray, optical, and infrared properties. A correlation between the X-ray power and both the optical continuum power and the permitted line widths was quickly established by Elvis and co-workers. Apparently there is a physical connection between the source of X-rays, the optical continuum, and the broad line spectra of type I Seyferts. The particular interest in this arises from the belief by astronomers that the optical emission is produced in the innermost nuclear region of the Seyfert galaxy, close to the ultimate source of energy. Continuing observations, mainly with *Ariel 5*, have confirmed this link and have also shown short-term X-ray variability which may relate directly to the dimensions of the 'powerhouse' in the nucleus of a type I Seyfert. Fig. 5.19 shows the typical X-ray flaring observed from NGC 4151 by *Ariel 5*.

The range of X-ray luminosity of the type I Seyferts is found to extend from NGC 3227 ($10^{35.2}$ W) to III Zw 2 ($10^{38.2}$ W), covering most of the thousandfold span between the prototype X-ray Seyfert (NGC 4151) and QSO (3C 273). Fig. 5.20 illustrates this fact and supports the view that Seyferts and QSOs lie on a natural physical sequence. Expressed another way, this suggests that the mysterious QSOs are (merely) more luminous

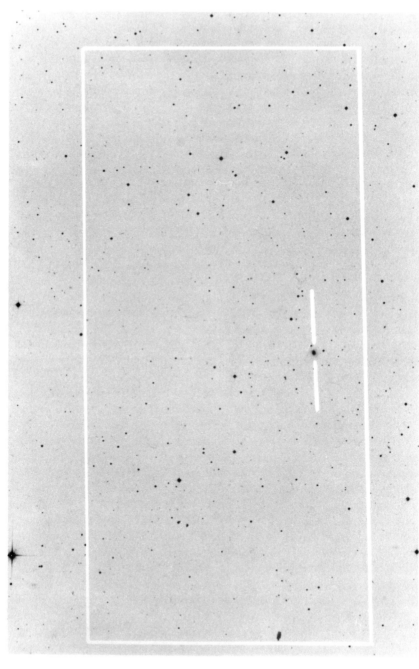

FIG. 5.17 X-ray error-box for the source 2A 0120-591 showing the bright galaxy ESO 113-IG 45 which optical spectroscopy showed to be a type I Seyfert of extreme luminosity.

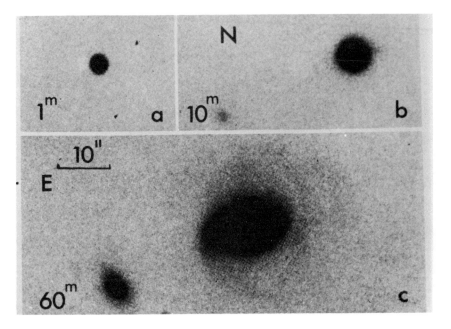

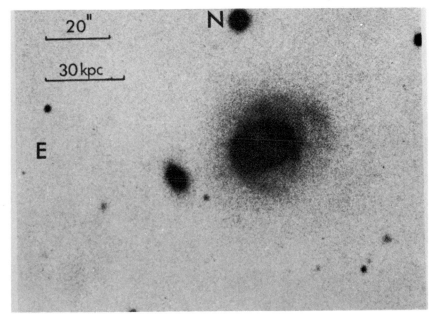

FIG. 5.18 Varying exposures of ESO 113-IG 45 showing the dominantly bright nucleus typical of Seyfert galaxies.

and distant Seyferts, where the surrounding galaxy has become too faint to see.

Following the successful work on the type I Seyfert galaxies, a number of less luminous active galaxies began to turn up in X-ray error-boxes. These, typically, also had much narrower emission lines and a lower X-ray luminosity, lying to the lower end of the Seyferts in Fig. 5.20. The probability that these emission line galaxies also have a compact central 'powerhouse' is strongly supported by recent observations of rapid variability in the X-ray emission from several such galaxies (Fig. 5.21).

The future

The future in X-ray astronomy is very near. It is, in fact, already orbiting the Earth in the shape of the Einstein Observatory (Fig. 5.22). This major new NASA spacecraft, an order of magnitude more massive (and more expensive) than *Ariel 5*, was launched on 13 November 1978 and carries a large, high resolution imaging X-ray telescope. The telescope, of the Wolter I design, employs two coaxial grazing incidence mirrors to produce an image of a small on-axis field with 3–4 arcsecs resolution. Combination of the large mirror (0.6 m) aperture and high resolution with the lengthy exposures possible in orbit gives the Einstein Observatory a sensitivity far beyond

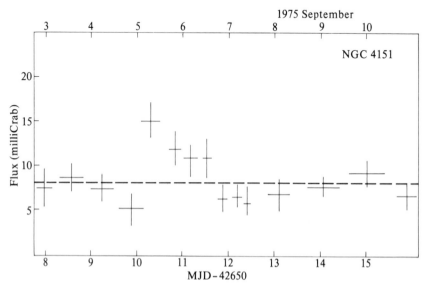

FIG. 5.19 Typical X-ray flare from NGC 4151 as seen in extended observations with the *Ariel 5* survey. Dashed line is mean flux level for 1974–8.

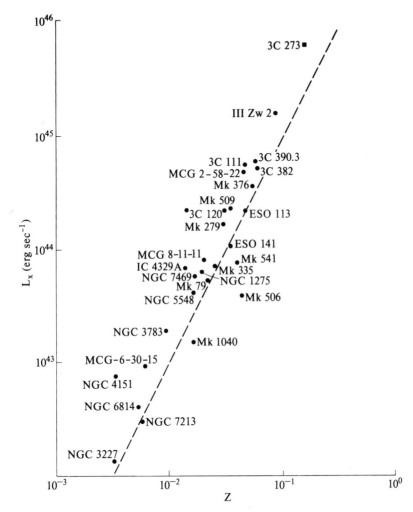

FIG. 5.20 Luminosity/red-shift diagram for Seyfert type I X-ray sources and for the QSO 3C 273. The dashed line shows, approximately, the sensitivity limit of the 2A Sky Survey at 0.5 *Ariel* count s^{-1}.

anything achieved hitherto. For example, a typical 10^4 s exposure will yield a point source detection a thousand times fainter than the limit of *Ariel 5*. Simple extrapolation of present results shows that active galaxies and clusters to red-shifts of $z \sim 1$ and beyond should be detected, in addition to many 'galactic sources' in other galaxies of the Local group and numerous low-luminosity stars. The extragalactic observations may well provide the first good evidence for the evolution of clusters, should discover many distant active galaxies, and could explain the mysterious 'X-ray background', detected in the very first Sco X-1 rocket flight, but still having contributions due to distant unresolved sources and a postulated diffuse 'background' emission which are quite uncertain. In addition to its remarkable sensitivity, the Einstein Observatory will provide the first images of cosmic X-ray sources with few arcsec resolution. This capability will be especially valuable for studying the X-ray distribution in supernova remnants, galaxy clusters, and other extended sources. Plates 4 and 5 show early examples of such data which promise over the next 2 to 3 years to greatly enhance our present view of the X-ray Universe.

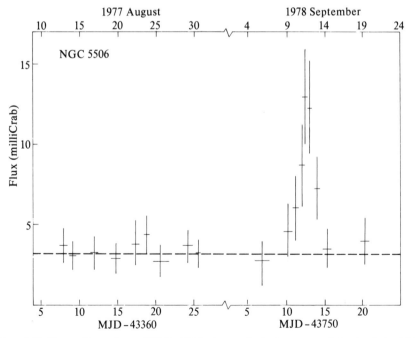

FIG. 5.21 X-ray flare from NGC 5506 detected during extended *Ariel 5* SSI observations. Dashed line is mean flux level for 1974–8.

(a)

(b)

FIG. 5.22 (a) Artist's impression of the Einstein Observatory launched on 13 November 1978. (b) Schematic diagram showing the Wolter I telescope mirrors and main instrumentation.

Bibliography

BAITY, L. A. and PETERSON, L. E. (ed.) (1979). *X-ray astronomy*. Pergamon Press, Oxford.

BLUMENTHAL, G. R. and TUCKER, W. H. (1974). Compact X-ray sources. *Ann. Rev. Astron. Astrophys.* **13**, 423–510.

GIACCONI, R. and GURSKY, H. (ed.) (1974). *X-ray astronomy*. D. Reidel Co., Dordrecht, Holland.

LEWIN, W. G. H. and VAN PARADIJS, J. (1979). What are X-ray bursters? *Sky and Telescope* **57**, 446–51.

OVERBYE, D. (1979). The X-ray eyes of Einstein. *Sky and Telescope* **57**, 527–34.

PETERSON, L. E. (1975). Instrumental techniques in X-ray astronomy. *Ann. Rev. Astron. Astrophys.* **13**, 423–510.

POUNDS, K. A. (1976). Rise and fall of an X-ray star. *New Scientist* **69**, 494–6.

—— (1978). An X-ray map of deep space. *New Scientist* **78**, 286–7.

—— (1979). Some recent results in X-ray astronomy. *Proc. R. Soc.* **366**, 277–489.

6

Black holes

ROGER PENROSE

If an object is projected in any direction away from the Earth's surface at a speed of more than about 7 miles per second then, frictional effects of the atmosphere apart, it will escape into space and not fall back to the surface. This is the speed, called the *escape velocity*, at which the kinetic energy of the object becomes equal to the (negative) potential energy of the Earth's gravitational field. At the surface of Jupiter the escape velocity is about 37 miles per second and at the surface of the Sun, about 135 miles per second. The general formula is

$$\text{escape velocity} = \sqrt{\frac{2GM}{R}}$$

where M is the mass of the gravitating body, R its radius, and G is Newton's constant of gravitation.

For bodies of fixed density, M is proportional to R^3, so the escape velocity gets larger (in proportion to R, or to $M^{\frac{1}{3}}$) as the mass gets larger. Indeed, it was noticed by Laplace* as long ago as 1798 that the escape velocity at the surface of a body of the density of the Earth and of radius some 250 times that of the Sun would be equal to the speed of light $c = 186\,000$ miles per second. Thus, Laplace inferred that such a body would be dark—in fact altogether invisible from large distances. The condition for the escape velocity to exceed c is

$$2GM > Rc^2$$

which, remarkably, is precisely the same as the modern 'relativistic' condition for a collapsed body to lie within a black hole. Laplace is therefore often credited with having effectively predicted the existence of black holes, namely situations in which a body has become so massive or concentrated that light is unable to escape from it.

* And also, somewhat earlier, by John Michell (1784).

However, Laplace's reasoning was, strictly speaking, fallacious. It was based on the Newtonian mechanics of his day, and in Newtonian mechanics it is not really possible to trap light in this way. The difficulty arises because in Newtonian theory there is no one absolute speed at which light must travel. Indeed, what *should* the speed of light be at the surface of Laplace's body? Imagine light emitted by a distant star falling towards the surface of that body. Suppose that the light strikes a horizontal mirror at rest on the surface. By the reversibility of Newton's laws, whatever speed that light has achieved on falling must be sufficient to lift it again away from the surface, so that it eventually reattains the distance of the far-away star and has the normal light speed c when it gets there. It seems unreasonable that light emitted at the surface of Laplace's body should be of an essentially different character from that falling on its surface, and that its speed be necessarily constrained in a way that Newton's theory does not demand. Thus, Laplace's Newtonian body should not be dark at all!

In relativity theory, however, the speed of light c does play an absolute and universal role. This suggests that some modified form of Laplace's argument might have some validity when the effects of relativity are taken into account. But they must be properly taken into account, according to Einstein's general theory of relativity. We shall see how to do this shortly. For the moment, let us just provisionally accept Laplace's relation $2GM > Rc^2$ as a criterion for a body to lie invisible within a black hole.

Is this criterion ever likely to be satisfied for realistic astrophysical bodies? In order to see that this is indeed to be expected, let us consider the evolution of an ordinary star like the Sun. Such stars obtain their energy by converting hydrogen into helium. Later, the helium itself gets converted into heavier elements. Then, at a certain stage, the density in the central region gets so great that the material there is converted to what is called *degenerate matter*, where, in effect, the electrons start getting in each other's way and the region is kept from collapsing because the Pauli exclusion principle of quantum mechanics forbids the different electrons from being in the same state. In this phase of the star's existence the outer parts of the star swell to enormous size and it becomes what is known as a 'red giant'. The Sun itself is expected to enter such a phase in about 10^{10} years when its surface will eventually reach out to the vicinity of the Earth's orbit or perhaps beyond. Then, as more and more of its material gets compressed into its degenerate central core, the Sun will contract inwards. Finally, its entire mass will become degenerate and the Sun will be a 'white dwarf', a star whose size is roughly that of the Earth itself.

But it was shown in 1931 by the Indian astrophysicist S. Chandrasekhar that there is a mass limit above which this kind of degenerate matter cannot be self-supporting—a limit which is only about one and a half times the mass of the Sun. Thus, while the evolution to a white dwarf star is possible for the Sun itself, the evolution of a star of, say, two or three times the Sun's mass cannot follow the same evolutionary path. At some stage, the growing degenerate core will exceed Chandrasekhar's limit and, consequently, collapse inwards. As the surrounding matter also falls in, having nothing to support it, it gets hot and compressed and indulges in violent nuclear reactions. Though there is some dispute among astrophysicists as to the details of what happens next, there seems little doubt that, at least in a large number of cases, the result is a supernova explosion—where, for a brief period of a few days a single star can outshine the entire galaxy in which it resides. Such an explosion could blow off the outer parts of the star, thus relieving it of perhaps most of its mass. This is not likely to save the core, however, whose inward collapse stops only when densities are reached, far in excess even of that of the electron-degenerate matter that the core originally consisted of (when a matchbox containing such matter would have had a mass of tons). The electrons have now become squeezed into the protons that inhabit the various nuclei that are present. This turns these protons to neutrons, and the neutrons themselves are squeezed together so that they get into each other's way, being saved from further collapse only by the Pauli principle again, which now prevents different *neutrons* from being in the same state. We arrive at a new kind of degenerate matter, called neutron-degenerate matter, in which the nuclei effectively congeal into one enormous atomic nucleus and the densities are comparable to that inside a nucleus itself. A matchbox of such material would have a mass of hundreds of millions of tons.

With the outer parts of the star either blown off or dragged into the core, the result would be what is known as a neutron star, with a total mass comparable with that of the Sun, but utterly tiny, being merely tens of kilometres across. Since 1967 such stars have been observed, either as the cores of pulsars or as certain X-ray sources.

But again there is a mass limit, referred to as the Landau–Oppenheimer–Volkov limit, above which a neutron star cannot support itself. There is some uncertainty as to the exact value of this limit, but a convincing upper bound can be placed at about three solar masses. (The actual limit must be somewhat less than this, though probably rather greater than the Chandrasekhar limit.) What, then, is the ultimate fate of a star whose initial mass is so large that, even allowing for substantial mass loss during the collapse phase, a core of more than three solar masses would be the result? Calculations show that although densities greater than that of a neutron star are attainable, no further stable self-supporting body can arise. Instead, the core collapses inwards until Laplace's 'black-hole' relation $2GM > Rc^2$ is achieved.

Before examining what this entails, it is worth while to note the vast range of star sizes over which standard astrophysical theory has worked well, and produced results in close agreement with observation. In Fig. 6.1 the comparative sizes of a red giant, the Sun, a white dwarf, and a neutron star are roughly indicated, providing a total range of a factor of some tens of millions in linear dimension and of more than 10^{20} in density. All the objects depicted have about the same mass, namely that of the Sun. It is perhaps remarkable that a further compression of only about a factor of five or so in linear dimension is needed in order that the black-hole condition be attained. So, from this point of view, black holes would seem to be quite probable objects. Stars are observed whose masses are 50 to 100 times that of the Sun, for example. Such stars would not be expected to throw off sufficient mass for a white dwarf or neutron star to result. Furthermore, the larger the mass of the resulting object, the larger would be the resulting black-hole radius in direct proportion to it, in accordance with the critical formula $2GM = Rc^2$, so the smaller would be the density

$$\rho_{bh} = M(\tfrac{4}{3}\pi R^3)^{-1} = \tfrac{3}{16}c^6\pi^{-1}G^{-3}M^{-2}$$

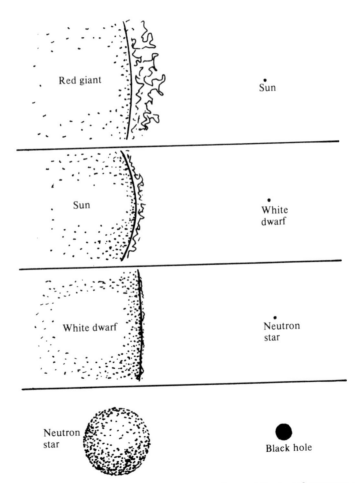

FIG. 6.1 Approximate relative sizes of red giant, Sun, white dwarf, neutron star, and black hole, all about one solar mass.

at which the body crosses over into the black-hole state. Thus, the larger the body, the less uncertainty there would be about the local physics at that stage. So, while it is true that a body of one solar mass would have to have densities in excess of the nuclear densities inside a neutron star as it collapsed into a black-hole state, this would not be so for very much larger bodies. Indeed, as Professor Rees has described (p. 31), it is thought that a black hole of some 5×10^9 solar masses may lie at the centre of the giant elliptical galaxy M87. For such a black hole, the radius would be about 10^{10} miles and the characteristic density ρ_{bh} would be only about one tenth of that of water. It should be borne in mind, however, that this is merely the density that a uniform body of that mass would attain just as it crosses through into the black-hole state. Vastly greater densities are to be expected subsequently, as the body collapses further inwards.

Such a large black hole would be expected to swallow gas and other stars. Velocities comparable with c would occur in its vicinity—and it is the presence of large velocities that lends weight to the belief that such a black hole indeed resides at the centre of M87. There is also a huge jet of luminous gas ejected from the centre, and it is believed that this may be the result of gas becoming highly compressed as it is sucked into the vicinity of the black hole, some of it being squirted out again along the rotation axis of the hole (which is assumed to be rotating).

It is quite probable that a black hole of about one million solar masses may lie even at the centre of our own Milky Way Galaxy. This would be about two million miles in radius. However, the observational evidence is here very scanty. The most convincing black-hole candidate, from the observational point of view, is the X-ray source Cygnus X-1, which would be a far smaller black hole of about ten solar masses. The black hole (if, indeed, this is what Cygnus X-1 is) is not directly seen, but X-rays are observed, as a result of gas being dragged into it and getting hot as the gas becomes highly compressed when nearing the vicinity of the hole. Neutron stars can also be X-ray sources, but in that case the signals tend to have a periodic intensity (owing to the rotation of the star's surface) as opposed to the irregular nature of Cygnus X-1. The main reason for believing that Cygnus X-1 is a black hole, however, is that it is a highly condensed object of about ten solar masses—much more mass than the theoretical maximum for a neutron star (or white dwarf). The estimate of its mass can be made because it is evidently in orbit around a companion star—a blue supergiant of known type and mass (about fifteen solar masses)—so that Newtonian dynamics provides a value for that of Cygnus X-1. The fact that Cygnus X-1 is a highly condensed object and not an ordinary star follows from the fact that it is an X-ray source. These two facts together provide the strong indirect evidence that it is indeed a black hole.

The picture is presented of these two objects in orbit about one another, material being dragged from the supergiant star to spiral about the black hole in a disc-like configuration, and finally to be swallowed by it (Fig. 6.2).

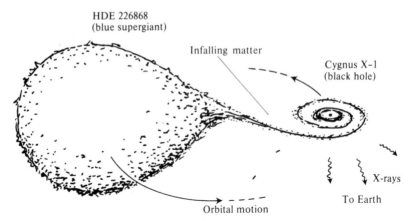

HDE 226868
(blue supergiant)

Infalling matter

Cygnus X-1
(black hole)

X-rays

To Earth

Orbital motion

FIG. 6.2 The standard black-hole picture of Cygnus X-1.

As the material spirals in, it gets more and more concentrated and hotter and hotter, until it reaches the temperature (about ten million °C) at which X-rays are produced, and then the material plunges into the hole.

Fig. 6.3 is a reproduction of a computer picture of the visual appearance of a black hole surrounded by such an accretion disc. The half-ring of illumination just above the black hole arises because the light rays close to the black hole are bent right around so that the accretion disc is seen again from its underside (Fig. 6.4). The calculations (by Luminet) were carried out on the basis of standard general relativity.

The evidence that Cygnus X-1 is itself a black hole is far from conclusive, however. It is just possible to provide rather contrived-looking models (involving triple-star systems) consistent with the observed data. So if one had strong reason for disbelieving in black holes, one would not regard Cygnus X-1 as compelling evidence for a black hole but presumably as evidence merely for such a 'contrived' arrangement. But the theoretical evidence that black holes *should* exist and, indeed, be fairly numerous objects in the Universe, does not rest on the interpretation of any individual object, such as Cygnus X-1. As was mentioned earlier, many stars are observed of mass up to one hundred times that of the Sun, there being no reason whatever to believe that enough mass will inevitably be ejected in the final collapse to produce a condensed object of below the Chandrasekhar or Landau–Oppenheimer–Volkov limit. Theory tells us that the resulting object must, instead, be a black hole and, on statistical grounds, one would expect a substantial number of black holes in our Galaxy. Similar arguments also apply even more convincingly to the central regions of large star clusters and, in particular, to centres of galaxies.

The essential ingredient in this theory, which applies in the final stage of the collapse to a black hole, is Einstein's general relativity. One must question, first, whether the observational evidence in favour of Einstein's

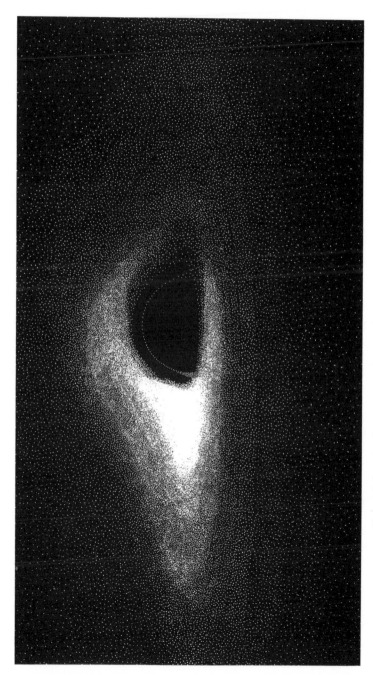

FIG. 6.3 Luminet's computer picture of a black hole with accretion disc.

theory is sufficiently impressive that its conclusions must be taken seriously. Secondly, one must ask whether black holes are indeed inevitable consequences of that theory.

This is not the place to enter into a detailed discussion of the observational evidence for Einstein's theory, but it may be said that this evidence is now very impressive. Most rival theories are convincingly disproved, the few that remain having been, for most part, contrived directly so as to fit in with those tests that have been actually performed. No rival theory comes close to general relativity in elegance or simplicity of assumption. The theory that

FIG. 6.4 The geometrical set-up assumed by Luminet for his calculations, showing how light bends around the hole, revealing the underside of the disc.

has been considered to be general relativity's most serious rival, namely the Jordan–Brans–Dicke theory, contains a parameter ω, the value $\omega = \infty$ providing a theory identical with general relativity. To be of astrophysical interest, a value as low as $\omega = 5$ was originally required. Observational evidence, from detailed examination of planetary motion within the solar system, has now placed a limit $\omega > 500$, for which values the theory would be indistinguishable from general relativity for all practical purposes. In any case, for *any* value of ω, this theory predicts black holes just as does general relativity.

The bending of light and radio waves by the Sun, an observed time-delay effect for such signals, the gravitational red-shift (clocks run slightly slow in a gravitational potential), and accurate tests of the principle of equivalence (cf. later) all lend considerable additional support to Einstein's theory. But perhaps the most impressive test yet is the observation of the binary pulsar, a double-star system consisting of a neutron star in (mutual) orbit about another condensed object, probably also a neutron star, for which a minute speeding up of the orbital period is observed, in perfect consistency with the prediction from Einstein's theory that such a system should lose energy in the form of gravitational waves. With all this evidence taken together, and with the fact that general relativity also inherits all of the successes of Newtonian gravitational theory, we must conclude that general relativity is, indeed, one of the most accurate theories known to science.

Let us, therefore, accept general relativity, not only as the best theory of gravitation available to us, but also as an excellent theory which indeed conforms closely with all the observed facts. What, then, is the nature of this theory? First of all, it is a theory of the four-dimensional geometry of space–

time, the principal ingredient of this geometry being its light-cone structure. A space–time point is a physical 'event', that is, something which has neither spatial nor temporal extent: an instantaneous pointlike occurrence. The light-cone of that event describes the history of a spherical flash of light which converges inwards to reach the event the instant it takes place, and then re-explodes outwards away from it. The history of the inward flash is called the 'past light-cone' of the event and that of the outward flash, the 'future light-cone' of the event (Fig. 6.5). Special relativity is described by a

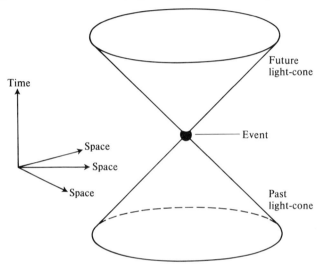

Fɪɢ. 6.5 The light-cone drawn in space–time.

geometry in which all these light-cones are uniformly arranged (Fig. 6.6), while in general relativity there may be substantial departures from such a uniform arrangement (Fig. 6.7). The normal conventions about diagrams depicting such space–times is that the future direction of time is represented as 'up' the page—although with gross departures from uniformity this rule can only be adhered to as a general tendency. The space–time description is the fundamental one in relativity, but it is often useful, for visualization purposes, to imagine the space–time to be divided up into a succession of slices, each of which is to be thought of as 'space' at 'one instant of time'. It should be emphasized, however, that such a slicing is quite arbitrary and has no physical status. It is made for convenience only, so that the space–time picture may be interpreted as a space evolving with time. If the slices are drawn 'horizontally' and if a light-cone is depicted in the standard way symmetrically up the page as is normal in special relativity, then the spatial picture is of a light flash symmetrically converging inwards and symmetrically diverging outwards, so that the light evidently has the same speed in all directions. But if the light cone is depicted as somewhat 'tipped over', then the speed appears to be greater in some directions than in others. This

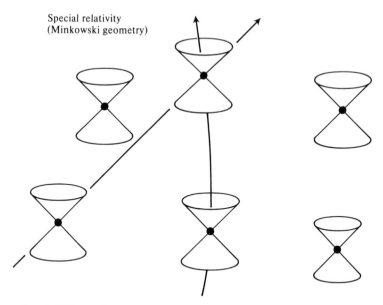

FIG. 6.6 The uniform arrangement of light-cones in special relativity.

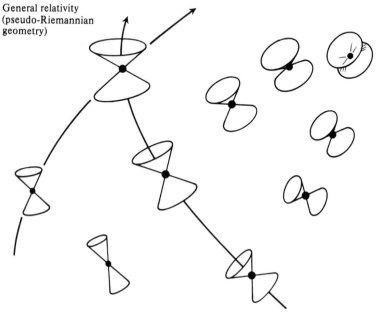

FIG. 6.7 Non-uniformly arranged light-cones in general relativity.

might, at first sight, seem to contradict the basic tenet of special relativity that asserts that the speed of light is always the same in all directions. However, this is not actually so. An actual measurement of the speed of light would, in fact, always yield the same result in every direction and give the value c. This is because the space–time structure in the immediate neighbourhood of any point is the same, in general relativity, as it is in special relativity. The peculiarities that seem to arise when a light-cone is depicted as 'tipped over' are a consequence only of the mode of description. (For example, with respect to a different family of space slices the cones need not appear to be 'tipped' at all.)

In some space–time diagrams a light-cone may be tipped so far as to have one side 'vertical', in which case the light in that direction would appear to hover in one spot, or it might be tipped even more, so that in the spatial description the sphere of light emitted by the event appears to move entirely off to one side! (Fig. 6.8). But again, this is not so paradoxical as it may seem. No observer may, according to relativistic causality, exceed the local light velocity. In the above instance this is interpreted to imply that an observer situated at the event in question must subsequently move off within the sphere of emitted light. It is not possible for him to 'stay in the same place' according to the given spatial description. This restriction on the possible motion of an observer follows from the fundamental requirement of relativity that physical massive particles have histories that are described, in the space–time picture, by curves (world-lines) that lie *within* the light cones of all events on the curve. Such curves are called 'timelike'. Light rays, on

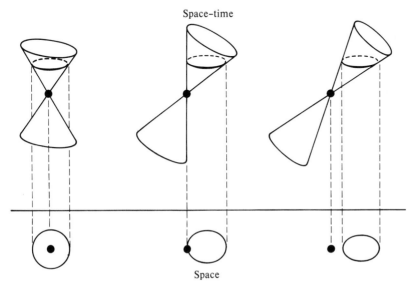

FIG. 6.8 Tipping light-cones.

the other hand, are curves that travel *along* the light-cones. Such curves are called 'null' (Fig. 6.9). The particles constituting the body of any physical observer must always be timelike curves. To such an observer, the 'tipping' of the light-cones that he encounters would not be observable at all.

Let us now consider a space–time picture which describes the gravitational collapse of a spherically symmetrical body to form a black hole, as originally suggested by Oppenheimer and Snyder. This is given in Fig. 6.10. The upper portions of the figure describe the stationary black hole and the lower portions, the collapse by which it was originally formed. The 'vertical' space–time 'cylinder', called the 'absolute event horizon', along which the light-cones are just so tipped that they have one side 'vertical' and tangential to it, is the boundary of the black hole. Inside the black hole we have the seemingly strange situation of light-cones tipped so far that it is impossible for an observer, or material particle, to 'stay in the same place' according to what would seem to be the most natural spatial description of the hole. In fact, any signal or material particle which is emitted inside the horizon is forced progressively inwards towards the centre and cannot escape outwards to be received by an observer at a large distance from the hole. It is clear, indeed, that the world-line of a light ray or particle cannot cross over from inside to outside the horizon, without violating the fundamental constraint of relativistic causality that was mentioned in the preceding paragraph.

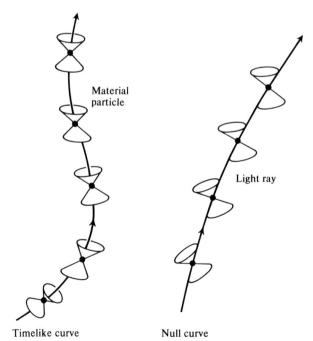

FIG. 6.9 World-lines of a massive particle and a photon.

Imagine an observer, at some distance from the hole, who looks straight at it. If, in the space–time picture, we trace a light ray from the observer's eye directly back in towards the hole, we find that as the horizon is approached, the light ray turns 'downwards' into the past, to be depicted more and more 'vertically', until the light appears in the remote past to be 'hovering' just outside the horizon. Then, as it is traced back further, the light ultimately encounters the collapsing material that first formed the hole. So the observer, no matter how long he waits, can, in principle, still see the matter, which appears always to be hovering just outside the horizon. However, in practice, he would very rapidly see nothing at all in the direction of the hole—just blackness. The reason for this is that only a very small finite portion of the history of the boundary of the collapsing body can be seen from outside, before that boundary is cut off by its crossing the horizon. The light from this finite portion of the boundary is all that can

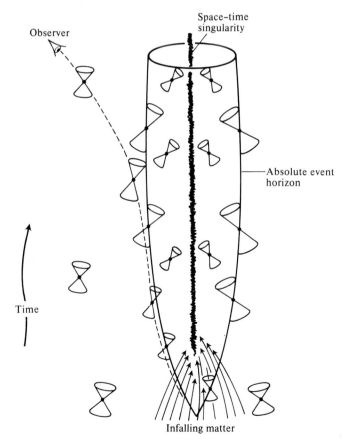

FIG. 6.10 Space–time picture of standard Oppenheimer–Snyder spherical collapse to black hole.

ever escape to be seen by external observers and it spreads itself over the entire infinite external time. Thus the apparent intensity of the body, as seen from large distances, dies down very rapidly (in fact exponentially) so that in a very short space of (external) time the intensity effectively drops to zero.

Suppose, on the other hand, we envisage an observer who (foolishly!) follows the matter inwards towards the centre. For him, nothing noteworthy happens as the horizon is crossed. Indeed, his only way of telling that this irrevocable occurrence had actually taken place would be to make some detailed calculations on the basis of the observations available to him. Only later, when the effects of space–time curvature begin to mount relentlessly and catastrophically—unaffected by any action that he might choose to take—would the full implications of his folly become apparent to him.

To understand more fully what is involved, in the notion 'space–time curvature', we must return to our discussion of general relativity. As was pointed out earlier, in a small enough region of space-time, there is nothing to distinguish general relativity from special relativity. The physical reason for this is Einstein's 'principle of equivalence' which tells us that gravitational force cannot be locally measured. It is, after all, the phenomenon of *gravitation* that distinguishes general from special relativity. So, the equivalence of the two theories on the local small-scale level rests on the fact that gravitational force cannot be locally detected.

But at first sight this seems an absurdity. Does not a simple spring-balance detect the gravitational force on the object being weighed? The difficulty here is that there is no local way of disentangling the effects of gravitational force from the effects of acceleration. If we drop the spring-balance rather than holding it still, then, on its way down, the balance will register *zero* for the apparent force on the object. Thus, to measure the gravitational force, one needs to know the acceleration of the measuring apparatus, and it is this acceleration that cannot be discerned locally (i.e. without reference to distant objects).

This indeterminacy stems from Galileo's observation that all objects, whatever their mass or constitution, necessarily fall at the same rate in a gravitational field. The apparent physical 'reason' behind this fact, in Newtonian gravitational terms, is the universal proportionality between passive gravitational mass (the response of objects to gravitation) and inertial mass (the resistance of an object to acceleration). This was roughly tested by Galileo's original experiments (such as his alleged dropping of unequal masses from the Leaning Tower of Pisa, and of other more accurate experiments that he certainly did perform), and it has now been verified to very great accuracy indeed by the successive experiments of Eötvös, Dicke, and Braginski. In Newtonian terms, this proportionality seems accidental, however, while in Einstein's theory it is elevated to a principle.

Since freely-falling masses drop at the same rate, the acceleration of each relative to any of the others is necessarily zero. In a freely-falling laboratory, the gravitational force is quite cancelled out. This is now a familiar aspect of space-travel. Astronauts float freely while in orbit, unhampered by any feeling of gravitational force whatever, even though they may be quite close to the Earth. The lack of apparent gravitational field results simply from the fact that everything falls together, following the same orbital path through space-time. So the astronauts feel zero gravitational force—an aspect of the principle of equivalence.

Yet it is not *quite* true that the gravitational field of the Earth cannot be locally detected. The problem arises with the meaning of the word 'locally'. How small a region must we be concerned with? And how accurately may we measure local accelerations? For the *deviations from uniformity* of the gravitational 'force' can indeed be measured. In Fig. 6.11 the situation is depicted of a particle falling freely in the Earth's field, surrounded by a sphere of other particles all of which are initially at rest with respect to the central particle. Then as the particles fall (or move freely together in orbit about the Earth), the pattern starts to deviate from spherical shape, and becomes deformed into an ellipsoid. From the Newtonian point of view, this distortion results from the fact that the gravitational acceleration is greatest nearest to the Earth, least farthest from the Earth, and slightly inwards for those particles at the same distance from the Earth as the central particle. This distortion to an ellipsoidal shape is called the 'tidal effect'. For if we replace the Earth in Fig. 6.11 by the Moon and the sphere of particles by the Earth, then we obtain the familiar pattern of tidal distortion of the Earth's oceans resulting from the gravitational effect of the Moon. From the point

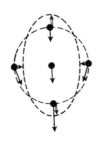

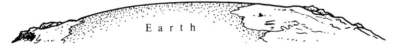

FIG. 6.11 Tidal distortion measures space–time curvature.

of view of Einstein's general relativity, this tidal effect represents the true and measurable gravitational field; it is the direct manifestation of space–time curvature.

The intensity of this tidal effect (i.e. of the curvature) varies as the inverse cube of the distance from the centre of a spherically symmetrical body. Thus, the effect is small for an observer at a large distance from a black hole. But as he falls inwards, the effect mounts—rapidly as the centre is approached. In the case of a black hole formed from ordinary stellar collapse, the curvature would already be easily sufficient to kill a man in free fall towards the hole, long before the horizon was reached. But for much larger holes, such as the one that apparently lies at the centre of the galaxy M87, he could approach and pass through the horizon without any noticeable effect whatever, since the curvature is still very small there. But then, as the observer falls in towards the centre, the curvature increases inexorably. Any physical object would finally be torn apart by the mounting tidal forces. It would break, first, into smaller pieces. Then its molecules would be torn from one another. Then the chemical bonds that hold the molecules together would be disrupted and the atomic nuclei separated from one another. Then the strong forces that bind the nuclei would succumb—and there is no reason to suppose that the elementary particles that compose those nuclei could themselves survive disruption when the curvature reaches so extreme a value that the radius of space–time curvature becomes less even than the radius of the particle. Indeed, if the exact (Schwarzschild) form of space–time metric that we have been discussing (and that Einstein's theory predicts for spherically symmetric collapse) holds true right down to the centre, then the intensity of the curvature reaches *infinity* there and the radius of curvature reaches zero! The centre is what is known as a 'space–time singularity', where the normal laws of physics break down.

But have we any right to believe that this exactly symmetrical model of a space–time is physically reliable, especially near the centre, where these enormous curvatures are anticipated? Any slight deviation from symmetry may be expected to get more and more magnified as the central regions are approached, and our exact model, based on the Schwarzschild metric, becomes less and less appropriate. Do we, indeed, have any reason to believe in the black-hole picture at all, except in the very special (and overwhelmingly improbable) particular case of exact spherical symmetry? To discuss this matter, we must now turn to the 'singularity theorems'.

What is first required is some qualitative criterion to indicate that irretrievable gravitational collapse has taken place, but one which does not depend on any assumption of symmetry. Several alternative such criteria exist. The simplest one to state asserts the existence of some space–time point P whose future light cone (swept out by the family of light rays into the future from P) begins to reconverge again somewhere to the future of P (Fig. 6.12). In spherical collapse, such points P occur near the centre in the late stages of the collapse. The reconverging of the light cone, as the rays pass through sufficiently concentrated matter, is one of the phenomena of general relativity that is, in effect, already *observed* in the bending of light by the Sun. The gravitational effect of the Sun acts as a converging lens. With enough matter in all directions surrounding the point P, this converging lens effect can be sufficient to cause the light rays in *all* directions out from P to begin to reconverge so that the criterion is satisfied. It is not required that this focusing effect be exact. It is a qualitative thing only, so that exact symmetry need not be assumed. The satisfaction of this criterion requires only that enough matter lie roughly concentrated in a small enough region, 'enough' here meaning that Laplace's condition $2GM > Rc^2$ is well satisfied—with something to spare so as to allow for the irregularities.

Assuming, then, that this criterion *is* satisfied, what do we conclude? One particular singularity theorem, due jointly to Stephen W. Hawking and the author, may be appealed to here. The remaining assumptions required for

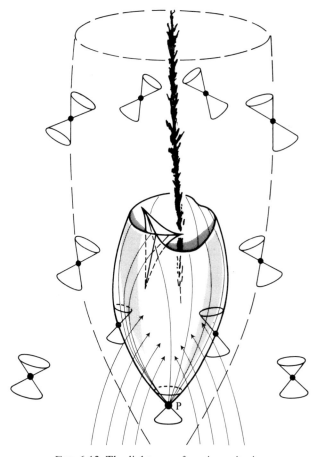

FIG. 6.12 The light-cone focusing criterion.

the theorem are that standard general relativity holds (without cosmological constant—or with negative cosmological constant, but either way this is unlikely to be an important restriction), that the pressure and energy densities satisfy some physically very reasonable inequalities ($\rho \geq 0$ and $\rho c^2 + p_1 + p_2 + p_3 \geq 0$, where ρ is the density and p_1, p_2, p_3 are the principal pressures), that there be no closed timelike curves (whose existence would imply the absurd possibility of an observer travelling into his own past), and that the space–time model be sufficiently general (a very mild and physically unexceptionable condition of curvature being not identically zero everywhere along timelike or null geodesics). From this, one concludes that a space–time singularity necessarily occurs. Thus the symmetry assumptions of the original Oppenheimer–Snyder collapse model are not necessary for the conclusion that a space-singularity must arise.

One is, of course, assuming that Einstein's space–time theory, and the normal ideas of the physical nature of matter, hold right down until the singularity is approached. But at some place, as space–time curvatures approach an absurdly large value, it is to be expected that the standard formalism of present-day physics must break down. This, indeed, is what must physically be meant by the term 'space–time singularity'—some as yet undiscovered physics must take over to govern the description of phenomena at such a 'singularity'. This is also the case for the space–time singularity of the Big Bang, which can again be shown, by means of singularity theorems, to be independent of any assumption of symmetry.

What is *not* so clear, however, as theoretical general relativity stands at the moment, is that the singularity resulting from gravitational collapse must inevitably be hidden from view by the presence of an absolute event horizon. With small deviations from spherical symmetry only, it seems fairly clear that the horizon will remain. But with a grossly different situation such as, for example, the result of two black holes spiralling into one another, it is merely *conjectured* that the resulting singularity will not be 'naked', and that it will necessarily be clothed by an event horizon to yield a final black hole. There is some moderately persuasive theoretical evidence for this conjecture, but it is yet far from conclusive. For the moment one frames it as a hypothesis—the so-called hypothesis of 'Cosmic Censorship', the proving of which remains perhaps the most important unsolved problem of classical general relativity. Most astrophysicists are happy to assume that the hypothesis is valid. With the only alternative to a black hole being, in effect, the much less acceptable 'naked singularity', Cosmic Censorship appears to be a very reasonable hypothesis!

With the assumption of Cosmic Censorship, one derives that a local gravitational collapse, in which the light-cone focusing criterion has been satisfied, *necessarily* results in a black hole. Since the collapsing matter need have no symmetry whatever, and may be very complicated in its mass distribution, it would seem, at first sight, that black holes also could have a very complicated structure, in general. However, this turns out *not* to be the case! As the black hole settles down to a stationary configuration, gravitational waves are emitted which carry away the irregularities in its field. According to a mathematical result (mainly the work of W. Israel, B. Carter, S. W. Hawking, and D. C. Robinson) the final stationary black hole is described by a particular space-time metric that had been discovered in 1963 by R. P. Kerr and which depends on only two parameters, describing the mass and angular momentum (spin) of the hole. This is a remarkable and very fortunate fact for black-hole theory, because it enables a surprisingly complete analysis of the structure of black holes to be carried out. This analysis is particularly simplified by a number of special mathematical properties that the Kerr solution turns out to possess. There is perhaps some irony in the fact that the astrophysical object that is strangest and least familiar to us, namely the black hole, should also be the one for which our theoretical description is the most complete!

One important property of black holes (which follows from the assumptions of Cosmic Censorship and non-negative energy densities) is the 'area increase principle'. This states that the surface area of a black hole's event horizon cannot decrease with time. As matter falls into the hole, this area increases. For a non-rotating hole, the area is proportional to the square of its mass, so the mass of the hole necessarily also increases. But if the hole is rotating the relation is a little more complicated and it is possible for the mass of the hole to decrease, provided that the spin is also reduced sufficiently. In effect, it is possible to extract mass-energy contained in the *rotational* energy of the hole. One might, for example, envisage a future civilization which obtains its energy in this way. A spaceship containing useless rubbish moves in a carefully selected orbit which enters a region close to a rotating black hole known as its 'ergosphere'. (The ergosphere is the region, lying just outside the event horizon, within which it is impossible for a material body to remain stationary as viewed by an observer at large distances. This is an example of the light-cone-tipping effect discussed earlier.) While in the ergosphere the spaceship ejects its rubbish into the hole in such a direction that it subtracts from the hole's total angular momentum and so that the spaceship escapes again to a large distance from the hole. It turns out that the kinetic energy of the spaceship can increase by even more than the total mass-energy of the rubbish! Some of the rotational energy of the black hole is carried away too (see Fig. 6.13). This is, in principle, a very efficient way of obtaining energy from a black hole, but whether a more realistic version of this process is likely to occur at all frequently in astrophysical situations remains unclear at present.

In the process just considered, an additional feature of note is the fact that the 'low grade' (i.e. high entropy) energy of the rubbish (which could have been simply 'lukewarm' radiation) has been traded for 'high grade' (i.e. low entropy) kinetic energy of the spaceship. At first sight it seems that the second law of thermodynamics (which states that entropy does not decrease) can be violated by processes of this kind. However, this does not take into

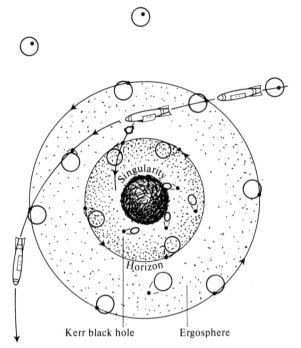

FIG. 6.13 Spatial picture of rotating black hole.

account the relentless and irreversible increase of the black hole's surface area. According to a careful analysis by J. D. Bekenstein, which takes into account the limitations of quantum mechanics, an entropy must be assigned to the black hole proportional to its surface area. Then the *total* entropy of the entire system (black hole and surrounding matter together) increases, and the second law of thermodynamics is saved.

Somewhat later, and using quite different methods (involving quantum field theory in curved background space–time) Hawking was able to improve on Bekenstein's result by showing that, according to quantum mechanics, a black hole ought also to *radiate*, at a temperature consistent with Bekenstein's formula. Hawking was also able to supply the precise value '$\frac{1}{4}$' for a proportionality constant that Bekenstein's methods had not been able to determine exactly, to yield the value

$$\frac{1}{4}\frac{kAc^3}{\hbar G}$$

for the entropy of a black hole of surface-area A. (Here k is Boltzmann's constant and $2\pi\hbar$ is Planck's constant.) Hawking's radiation is a necessary requirement for the consistency of the whole scheme. It arises, roughly speaking, because virtual particle–antiparticle pairs arise out of quantum fluctuations, and, in the neighbourhood of the hole, these can become real pairs, one member of each pair being swallowed by the hole, the other escaping to infinity (Fig. 6.14). This flux of quantum particles escaping to infinity has a thermal ('black body') nature and constitutes the Hawking radiation.

For a non-rotating hole of mass M, the value of this temperature is

$$\frac{hc^3}{8\pi kGM},$$

which for any black hole that could be formed from stellar collapse is absurdly small (less than 10^{-7} K) and is quite undetectable. However, Hawking has speculated that very tiny black holes might have been formed in the Big Bang, and, if small enough, the temperature could become significant. For a hole formed with as little mass as 10^{15} g (whose radius would then be less than the classical electron radius) this temperature would be high enough that by the present age of the Universe the hole could have just about evaporated away its entire mass in the form of Hawking radiation. In fact, such a hole would be expected to have an explosive end since as it grows smaller it gets hotter, its energy being emitted more and more rapidly towards the end. The final explosion would release about 10^{30}–10^{35} ergs which, though large by terrestrial nuclear weapon standards, is still small when compared with many astrophysical explosive events. Such 'mini-hole' explosions have been looked for, but none have been seen.

In principle, even a large black hole would be expected eventually to decay away in the same way, but the time-scales involved are ridiculously long. For example, a black hole of one solar mass would require 10^{53} times the present age of the Universe for this to take place! Possibly the Universe will recollapse long before this happens!

In my own opinion mini-holes are unlikely to exist. For reasons connected with the second law of thermodynamics it seems that the Big Bang singularity must have been very 'uniform' in nature. Indeed this accords well with various different cosmological observations. A 'chaotic' Big Bang would have been necessary for mini-holes to form.

This uniform nature of the Big-Bang singularity contrasts markedly with the expected chaotic nature of the singularities inside black holes. This is connected with the black holes' singularities having a much higher entropy. It even seems that there is a law of nature (as yet undiscovered) that forced the Big Bang's singularity to be strongly constrained in a way that does not apply to the final singularities of collapse. An understanding of why this should be, and of the exact nature of this constraint, may eventually supply an understanding of the origin of the second law of thermodynamics and of many important unanswered questions of cosmology. Thus, future physical theory is presented with many profound and far-reaching challenges in discovering the laws that govern space–time singularities!

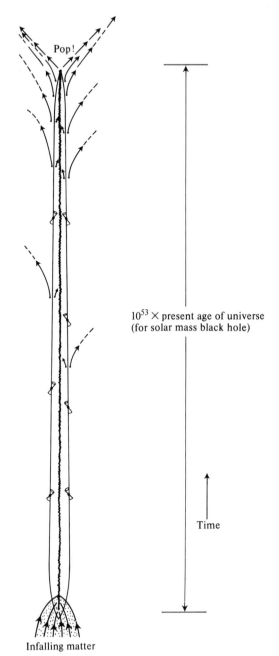

FIG. 6.14 Hawking evaporation of solar mass black hole.

Bibliography

Non-technical

KAUFMANN, W. J. (1977). *Cosmic frontiers of general relativity.* Little, Brown, Boston and Penguin Books, Harmondsworth.
PENROSE, R. (1973). Black holes. In *Cosmology now* (ed. L. H. John). BBC publication, Broadwater Press, Ltd. Welwyn Garden City.
WALD, R. M. (1977). *Space, time and gravity: the theory of the Big Bang and black holes.* University of Chicago Press.

Technical

BEKENSTEIN, J. D. (1973). *Phys. Rev.* **D7**, 2333.
—— (1974). *Phys Rev. D9*, 3292.
HAWKING, S. W. (1975). *Commun. Math. Phys.* **43**, 199.
—— and ELLIS, G. F. R. (1973). *The large scale structure of space–time.* Cambridge University Press.
—— and Israel W. (eds) (1979), *General relativity: an Einstein centenary survey.* Cambridge University Press.
MISNER, C. W., THORNE, K. S., and WHEELER, J. A. (1973). *Gravitation.* Freeman, San Francisco.
PENROSE, R. (1969). *Rivista del Nuovo Cimento* Ser. I. Vol. 1, Numero speciale 252.

7

Planetary exploration

GARRY E. HUNT

Introduction

For centuries discussions about the origin of the solar system and the origin of life were no more than idle speculations using theories thinly supported by limited scientific evidence gained from telescope observations. With the beginning of direct exploration of the solar system, planetary science has rapidly developed into one of the most active areas of study. Planetary scientists are now confronted with such a huge number of new observations and theories that a revolution in understanding the solar system is in progress.

Until recently planetary studies were regarded as part of astronomy. However, the advent of space probes in the 1960s provided new tools for detailed planetary investigations, as a result of which the study of planetary meteorology, aeronomy, magnetospheres, geophysics, geology, and exo- biology rapidly emerged. These involved, in the main, research workers from the parallel earth-science disciplines. Earth and planetary studies thus tended to merge with one another, whilst planetary science tended to diverge from astronomy. From an academic point of view, earth-science studies have now expanded to include the whole of the solar system. At the same time the rapid increase in our knowledge of the individual members of the solar system has given rise to the new science of comparative planetology.

These developments are of fundamental importance for all scientists. Consider, for example, the geological and meteorological properties of the planets. The planetary surfaces reflect the basic characteristics of the body, the properties of its interior, volcanism, and external effects such as bombardment by meteorites and other fragments of material. Dating the resulting surface features is the first task that must be tackled if we are ever to learn the time-scale of events that have shaped the surfaces. The early solar system, it seems, was anything but a peaceful, quiescent place. The record of bombardment contained in the cratered surface of the Moon establishes that the flux of material was so high during the first 600 million years of lunar history that it saturated the original crust with crater upon crater, and the flux decreased very rapidly between 4 and 3.3 billion years ago and may have remained nearly constant since. Apparently similar cratering histories have now been observed on Mercury and Mars, lending support to the idea that sustained bombardment was a phenomenon common to the entire *inner* solar system. What can we then expect to find on the surface of Venus beneath the ubiquitous cloud layers? How old is the atmosphere, and how has it affected the surface characteristics? These are fundamental questions that remain to be answered. We now know that some of the *outer* solar system bodies also show the effects of a heavy bombardment. Callisto is found to be the most densely cratered object in the solar system. Obviously, we must determine the time-scale for these events and investigate whether they occurred at roughly the same time as the bombardment of the inner-solar-system planets.

Nowhere is comparative planetology more important than in studies of planetary atmospheres. Knowledge of the atmospheres of planets provides fundamental information about their origin and evolution, thereby providing important constraints on theories of the observed distribution of material in the solar system. This information will enable a quantitative investigation of whether these alien atmospheres could support our own or other forms of life.

In addition, there are the vital studies in planetary meteorology. The variation in atmospheric conditions that constitute weather and climate on Earth are of major scientific, economic, and political importance, particularly on the time-scale of a human lifetime. But our understanding of these phenomena is far from complete. We do not know whether man's activities have any effect or whether external effects such as solar activity can modify the climate. Since terrestrial meteorology is complicated by the effects of oceans, continents, mountains, and clouds, some perspective is emerging from studies of atmospheric phenomena on the other planets.

Spacecraft missions to the planets have only occurred during the last 17 years, but they have brought an explosion in our knowledge of these tiny specks of light in the night sky since they can now be observed at resolutions of one kilometre or less (see Table 7.1). All the planets from Mercury to Jupiter have now been observed at close range, and, before the end of 1979, we will obtain the first glimpse of Saturn with the *Pioneer* spacecraft, followed by detailed investigations by *Voyager* in 1980 and 1981. In this chapter, I briefly review current understanding of these bodies, leaning heavily toward comparative planetary processes, and the challenging problems of the next decade—a subject which is both exciting and scientifically rewarding.

Mercury

As Copernicus lay on his deathbed he lamented never having seen the planet Mercury. Since Mercury is the innermost planet in the solar system, and never more than 23 degrees from the Sun in the sky, it is notoriously difficult to study by conventional techniques. Now, more than 500 years after the founder of modern astronomy was born, the spacecraft *Mariner 10* (see Table 7.1) has passed within a few hundred kilometres of Mercury and

Table 7.1
Planetary spacecraft missions

Planet	Spacecraft	Launch	Mission summary
Mercury	*Mariner 10*	3 Nov. 1973	Flyby 29 March 1974, Sept. 1974, March 1975
Venus	*Mariner 2*	27 Aug. 1962	Flyby 14 Dec. 1962
	Venera 3	16 Nov. 1965	Landed 1 March 1966
	Mariner 5	14 June 1967	Flyby 19 October 1967
	Venera 5	5 Jan. 1969	Atmospheric entry 16 May 1969
	Venera 6	10 Jan. 1969	Atmospheric entry 17 May 1969
	Venera 7	17 Aug. 1970	Soft surface landing 15 Dec. 1970
	Venera 8	27 March 1972	Soft surface landing 22 July 1972
	Mariner 10	3 Nov. 1973	Flyby 5 February 1974
	Venera 9	8 June 1975	Orbiter and soft lander 22 Oct. 1975
	Venera 10	14 June 1975	Orbiter and soft lander 25 Oct. 1975
	Pioneer Venus 1	20 May 1978	Orbiter, 4 Dec. 1978
	Pioneer Venus 2	8 Aug. 1978	Multi probes—atmospheric entry 9 December 1978
	Venera 11	9 Sept. 1978	Soft surface landing 25 Dec. 1978
	Venera 12	14 Sept. 1978	Soft surface landing 21 Dec. 1978
Mars	*Mariner 4*	28 Nov. 1964	Flyby 14 July 1965
	Mariner 6	25 Feb. 1969	Flyby 31 July 1969
	Mariner 7	27 March 1969	Flyby 5 August 1969
	Mars 2	19 May 1971	Orbiter: hard landing probe 27 Nov. 1971
	Mars 3	28 May 1971	Orbiter: soft landing probe 2 Dec. 1971
	Mariner 9	30 May 1971	Orbiter—13 November 1971
	Viking 1	20 Aug. 1975	Orbiter: soft landing 20 July 1976
	Viking 2	9 Sept. 1975	Orbiter: soft landing 3 Sept. 1976
Jupiter	*Pioneer 10*	3 March 1972	Flyby 4 December 1973
	Pioneer 11	6 April 1973	Flyby 3 December 1974
	Voyager 1	5 Sept. 1977	Flyby 5 March 1979
	Voyager 2	20 Aug. 1977	Flyby 9 July 1979
Saturn	*Pioneer 11*	6 April 1973	Flyby 1 Sept. 1979
	Voyager 1	5 Sept. 1977	Flyby approximately 12 Nov. 1980
	Voyager 2	20 Aug. 1972	Flyby approximately 20 Aug. 1981
Uranus	*Voyager 2*	20 Aug. 1977	Possible Flyby around Jan. 1986
Neptune	*Voyager 2*	20 Aug. 1977	Possible Flyby around Sept. 1989

provided high-resolution images of the surface and surprising discoveries of the neighbouring environment.

Mercury has a rotation period of approximately 59 days compared with the 88-day orbital period of the Mercurian year. This means that there is a 2:3 ratio of the rotational and orbital periods so that the planet rotates exactly three times while circling the Sun twice. This precise relationship was demonstrated by the *Mariner 10* observations since the spacecraft encountered the planet three times at intervals of 176 days (see Table 7.1). Mercury has the harshest surface environment of any planet in the solar system with equatorial temperatures varying from 700 K at noon to 100 K at night.

The planet possesses two distinct hemispheres, as shown in the *Mariner 10* images in Fig. 7.1. The left hemisphere shows a variety of cratered terrain which resembles the heavily cratered regions of the Moon, with the crater Kuiper prominent. This crater is approximately 40 km in diameter and reflects sunlight more strongly than any other feature on the planet. However, as a whole, Mercury is a very dark object, with an albedo of only about 6 per cent. The right hemisphere shows the structure of the large 1400 km Caloris basin, comparable in size to the Imbrium basin on the

FIG. 7.1 The hemispheres of Mercury photographed by *Mariner 10*.

Moon, and probably created by a large impact. Fractures have been detected in the floor of this basin together with 'young' craters 35 km in diameter. The Caloris basin appears to be filled with smooth plains material.

In a general way the surface of Mercury closely resembles the Moon. There are, however, important differences revealed by the *Mariner 10* observations. Unlike the Moon the surface of Mercury is not saturated with large craters with diameters in the range 20 to 50 km. The force of gravity on Mercury, which is twice that of the Moon, affects the structure of the cratered terrain. Material ejected from a similar primary impact on Mercury covers an area only a sixth as large as the area covered on the Moon, while secondary craters on Mercury are much more closely clustered around primary craters. A further important difference between the heavily cratered regions of Mercury and those of the Moon is the presence throughout Mercury of shallowly scalloped cliffs running for hundreds of kilometres. These lobate scarps may have resulted from an early period of crustal compression on a global scale, produced by the slow cooling and contraction of Mercury's large iron core. Mercury exhibits large, well preserved craters, which are probably three to four billion years old, and this confirms that there has been no Earth-like migration of crustal plates since that time. Furthermore the lack of surface erosion rules out any tangible atmosphere on Mercury. In contrast, the eroded surface of Mars shows how even a tenuous atmosphere quickly modifies the appearance of large craters.

While the surface of Mercury closely resembles the Moon, the interior of the planet is more like the Earth. Mercury has a weak dipole magnetic field whose strength ranges from about 350 to 700 gammas at the surface, which is approximately one per cent of the strength of the Earth's field. The Mercurian field is unexpectedly aligned with the spin axis. This is an important discovery. Although Mercury certainly has a large iron core, the rotation of the planet is so slow that one would not expect a dynamo mechanism to be capable of producing this magnetic field. Clearly, the origin of the field remains a major unresolved problem. Its solution will require a deeper understanding of the mechanism of the Earth's field to find a process that will allow it to be reduced to a Mercurian scale.

The *Mariner 10* mission to Mercury completed the reconnaissance of the inner solar system and once more demonstrated that planetary exploration is full of surprises. The next step in a planned programme of exploration would be an orbiting spacecraft. It would require instrumentation to map portions of the surface not yet observed, to measure the solar wind interaction, and to study those magnetic field properties which we do not yet understand.

Venus

Venus is the brightest object in the sky after the Sun and Moon and is our nearest planetary neighbour. It has a mass and radius which are very similar to those of the Earth. At first sight this would suggest that Venus is our planetary sister. But there the similarity ceases, for the atmospheric composition and meteorology of Venus are very different from their terrestrial counterparts.

Throughout the centuries, numerous photographs of Venus have produced only one frustrating result; the planet is covered by a uniform unbroken layer (or layers) of yellowish cloud whose top may extend up to 100 km above the surface. This cloud-cover reflects 79 per cent of the total sunlight it receives and consequently Venus has the highest planetary albedo in the solar system. This is in complete contrast to the Earth which has an average 50 per cent cloud cover with a planetary albedo of approximately 30 per cent.

Atmospheric composition

The Venus atmosphere is huge, more than 90 times more massive than the Earth's and is composed primarily of carbon dioxide. Traces of hydrochloric and hydrofluoric acids and carbon monoxide have been detected, together with very small amounts of water-vapour. But the real surprises have come from the recent measurements made by the *Pioneer* Venus probes as they descended through the atmosphere. The observations show large concentrations of argon and neon—^{36}Ar, ^{38}Ar, and ^{20}Ne. The volume mixing ratio of ^{36}Ar in the Venus atmosphere is approximately 10^{-4} compared with a value of 3.2×10^{-5} for the Earth. Since the atmosphere of Venus is approximately 90 times more massive than that of the Earth, it follows that the absolute abundance of ^{36}Ar in the Venus atmosphere must be approximately 200 to 300 times larger than on Earth. The abundance of ^{20}Ne is also higher in the Venus atmosphere and comparable to that of ^{36}Ar. The ratio ^{36}Ar:^{38}Ar for the Venus atmosphere is similar to values observed for the Earth, meteorites, and the Moon.

The ^{36}Ar measurement has a profound effect upon theories of the formation of the solar system. Combined with the *Viking* measurements for Mars, we find that the amount of argon decreases from Venus to Mars. We now believe that the solar nebula, out of which the planets were formed, had an inhomogenous element distribution. The inert gases, such as argon, were probably trapped in the grains that formed the planets and formed a veneer on these particles as they condensed out of the nebula.

Surface properties

The Russian *Venera* and the US *Pioneer* probes have measured the surface-temperature of the planet to be approximately 737 K, which is more than three times the highest temperature measured on the surface of the Earth. It also appears that the surface-temperature of Venus does not vary greatly with either latitude or solar phase angle, or diurnally. Since Venus has a very slow rotation period of 243 Earth-days and a long day (120 days) the planet's motion is inefficient in transporting heat and reducing temperature contrasts that would otherwise be suppressed by differential solar heating. The slow rotation of the planet may also account for the absence of any detectable magnetic field.

Although the surface of Venus is hidden beneath the clouds, we are gradually beginning to discover some of the 'hidden' mysteries of this region. Radar images from ground-based observations have revealed a heavily cratered surface with depressions ranging in size from several hundred km to a resolvable limit of about 30 km. The *Pioneer* Venus measurements have scanned only a limited portion of the surface, and disclose a rift valley having relief up to 7 km as well as gently rolling plains. This may be interpreted as evidence, now or in the past, of tectonic activity and volcanoes. The *Pioneer* Venus radar data is certainly consistent with current ideas that volcanic activity played an important role in the development of the atmosphere on Venus.

The atmosphere will affect the surface domain. The images obtained by the *Venera 9* and *Venera 10* spacecraft display a rock-strewn terrain. At a pressure of about 90 atm, and temperature of 747 K, surface wind speeds of 1 m s^{-1} have been recorded. Erosion rates may be very slow on Venus, since the blocks seen in the Venera images have sharp edges. There seems to be some fine dust on the surface. The Pioneer Venus 'day probe' landed on a surface of loosely compacted material and a small amount of fine dust was ejected into the atmosphere in the vicinity of the probe. The dust subsequently settled on to the surface over the next few minutes until the surrounding atmosphere was again free of particulate matter.

Mysterious observations were made both by the *Pioneer* Venus and *Venera 11* and *Venera 12* spacecraft during the last minutes of their descent to the surface. They suggest the occurrence of lightning at lower altitudes where *Venera 12* counted at least 1000 impulses of radio noise in the altitude range 8–14 km. Venusian lightning is not unreasonable, since the atmosphere has been found to be electrically active. By way of comparison, there are typically 100 lightning strokes every second scattered all over the Earth, whereas on Venus there may be several times that number in a localized area. Instead of being illuminated for a brief instant by a dazzling flash of lightning, as on the Earth, the Venusian sky may well be glowing from the nearly continuous discharge of frequent lightning strokes.

Meteorology

Although the visible appearance of Venus is a featureless disc, observations in ultraviolet light have provided a most exciting and unexpected result. In the neighbourhood of the cloud tops the upper atmosphere super-rotates. Dark features in the shape of a Y or C move in a retrograde manner with a rotational period of only 4 days, implying zonal winds of about 100 m s^{-1}. This rotational period is in the same direction as the planet's rotation but is very short when compared with the solid-body rotation of 243 days. Venus possesses the largest ratio between atmospheric and planetary rotation-rates of any body in the solar system.

What mechanism or mechanisms are capable of generating these huge winds, which are in the opposite direction to, and 20 times faster than the overhead motion of the Sun relative to a fixed point on the Venus surface? Almost all the solar energy is absorbed in the neighbourhood of the cloud tops and provides the heating which drives the observed motions. Its effects are conveyed to higher levels by transport processes. In the stable upper atmosphere of Venus, internal gravity waves are expected to play a major role in the vertical transport of heat. Since these mechanisms do not act simultaneously the horizontal movement of the source of heat produces a tilt in the vertical pattern of convection which results in a net motion in the opposite direction to that of the Sun. It is possible to create mean motions much faster than the speed of the heat source. The magnification factor is largely determined by the deviation of the vertical lapse rate of the temperature from the adiabatic value and the distribution of radiative heating and cooling in the Venusian atmosphere.

The high resolution images of Venus obtained during the *Mariner 10* flyby (see for example Fig. 7.2) show many important small-scale features. The maximum solar heating occurs at the subsolar point characterized in the picture by cellular features typical of condensation processes. The interaction of the zonal winds with the subsolar disturbance give rise to bow-shaped waves. In the equatorial zones outside this disturbed region, light and dark features move westward with velocities of about 100 m s^{-1}. In other regions there is a more complicated picture of an increase in angular velocity with distance from the equator up to a latitude of approximately 50 degrees, where the rotational period may be as short as two days. Spiral streaks which may be interpreted as 'jet streams' show a flow toward the polar vortices where the kinetic energy of the atmospheric motion is eventually dissipated. Many of these features can be seen in subsequent *Pioneer* images, although five years later a significant change is seen in the polar regions. Then the poles are more uniformly covered by clouds to 50 degrees latitude (Plate 6).

To maintain the planetary angular momentum balance the return flow of atmospheric gas probably occurs at deeper levels in the atmosphere. But we do not know the vertical extent of these motions; whether the stratospheric rotation is a separate overturning cell or whether it is linked somehow to a deep stirring of the huge Venusian atmosphere. Certainly there will be a rapid reduction in the zonal winds as we descend through the troposphere, and the Venera spacecraft have measured values of only a few metres per second in the lowest layers of the atmosphere.

But why is the meteorology of the upper atmosphere of Venus so different from that of the Earth? The super-rotation requires an efficient high altitude thermal forcing, but for the Earth the largest amount of available radiative energy is that absorbed at the surface. Could other planetary atmospheres

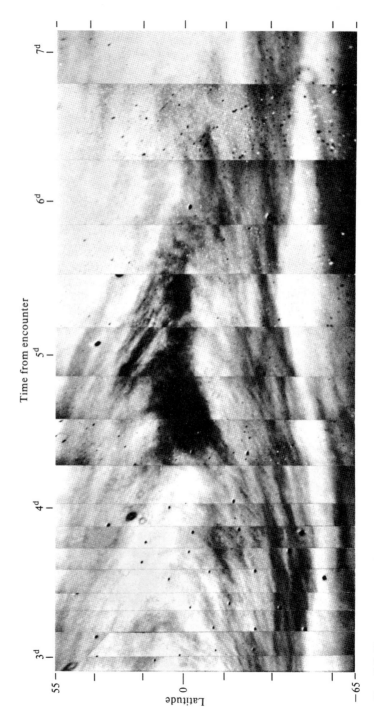

FIG. 7.2 The appearance of the UV markings as they crossed the subsolar meridian of Venus during the period 2.9 to 7.1 days after encounter. The images have been radiometrically decalibrated. (After Anderson *et al.* (1978). *Ap. J. Supp.* Series 36, 275–84.)

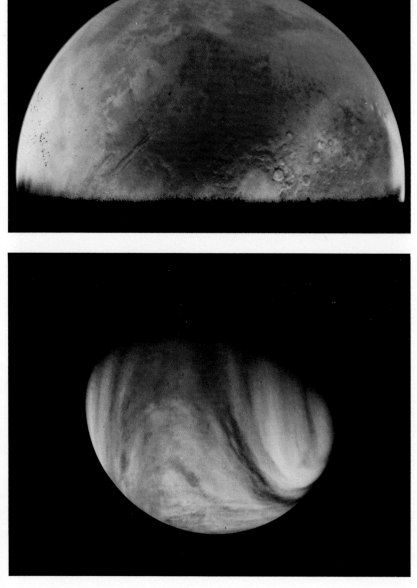

VI. A UV image of Venus obtained on January 14 1979, by the cloud photopolarimeter aboard the Venus *Orbiter* spacecraft. The phase angle of the centre of the disc is 57°.

VII. View of Mars from *Viking*. Just below the centre of the image and near the morning terminator is the large impact basin Argyre. North of this region is the extensive canyon Valles Marineris.

VIII. Jupiter as observed by *Voyager 1* on 29 January 1979, showing the Great Red Spot.

IX. High resolution image of the Great Red Spot obtained by *Voyager 1* on 1 March 1979. A white oval is visible in the neighbourhood of the Great Red Spot. Both features show a turbulent flow to the west of their location.

exhibit this excess rotation? Certainly any atmosphere where heating tends to produce an equatorial thermal bulge, but which cannot develop instabilities because of slow rotation or large damping, must develop an excess rotation at a high level in the same sense as the planetary rotation.

The composition of the Venus atmosphere presents several problems. Although oxygen is a major component of the Earth's atmosphere, its absence from Venus presents an enigma. The carbon monoxide of Venus is presumably formed in the upper atmosphere when CO_2 is dissociated by ultraviolet sunlight. Oxygen is an inevitable byproduct of this reaction and since it is a diatomic molecule we may expect half as many O_2 molecules as CO molecules. This is indeed the case on Mars, which also has a predominantly CO_2 atmosphere, but on Venus oxygen is 50 times less abundant than carbon monoxide. We could explain this deficiency by requiring the rapid transport of oxygen to the lower atmosphere where it could combine with other substances, such as sulphur. If this is the case, mixing of gases in the Venus stratosphere would have to be much more efficient than it is on the Earth or on Mars.

A further major problem is the apparent dryness of the Venus atmosphere. In the neighbourhood of the cloud tops the relative humidity rarely reaches one per cent, while in the lower atmosphere only 0.1 to 0.2 per cent water-vapour has been detected.

Atmospheric structure

Both of these problems are related to the composition of Venus's clouds, whose yellow colour has long been cited as evidence that, unlike the white, terrestrial clouds on Earth, they do not consist of water or ice. Detailed analysis of the polarization of reflected sunlight from the clouds is consistent with concentrated sulphuric-acid droplets. Certainly the hygroscopic nature of this substance will account for the very low water-vapour concentrations detected in the neighbourhood of the cloud tops, but the lemon colour of the cloud would still remain unexplained. The presence of elemental sulphur is thought to be responsible, although it is not ordinarily the same colour as the planet. However, this objection is not compelling, as the colour of sulphur is strongly dependent upon the local temperature.

The most definite information on Venus's clouds has come from the recent *Pioneer* measurements which show a haze layer overlying the main cloud deck. This haze is the outer cloud layer observed by ground-based observers, and was first identified as being composed of sulphuric acid droplets of radius between 1–2 μm.

The middle cloud region between 51 and 56 km shows a decrease in concentration of particles to about $100\ cm^{-3}$, but involves larger liquid droplets and solid grains of radius 10–20 μm. The densest layer lies between 49 and 52 km where visibility is restricted to less than 1 km. This is the region with the largest particle sizes. Just beneath this opaque layer is a distinct thin region of somewhat reduced opacity.

In the region from 32 to 48 km lies a thin haze of particles nearly 1 μm in size with concentrations of 1 to 20 cm^{-3}. The upper part from 45 to 47 km is most dense. Below the lower boundary the atmosphere appears free of particles all the way to the surface. However, this does not mean that the visibility is unlimited. Venus's atmosphere is so dense that even in the absence of clouds the scattering of light by gas molecules alone would probably hide the surface from view.

In general the measurements indicate a concentration of particles either independent of height or decreasing at lower altitudes. This behaviour is consistent with a source at the top and a sink at the bottom with downward flow by mixing and by gravitational settling. This suggests that there is a source of sulphur and sulphuric acid at the top of the middle cloud region at about 57 km. The source is presumably photochemical in nature, requiring ultraviolet radiation that does not penetrate far into the sulphur cloud. The H_2SO_4 droplets lose water on the way down, but then encounter increased H_2O, and perhaps SO_2 and SO_3 as they fall under gravity to form a layer of dilute acid which is the densest layer in the lower cloud region. Most importantly, the large sizes (greater than 10 μm) encourage the possibility of precipitation on Venus, either from the dilute H_2SO_4 lower cloud region, or initiated by sulphur from the middle cloud region which would pick up dilute H_2SO_4 when falling below 51 km.

Besides water, several substances are almost non-existent above the clouds, but well represented below them. The abundance of SO_2 varies in this way by a factor of 240 and oxygen by 60. These higher abundances arise when sulphuric acid and sulphur rain out of the clouds, vapourize and even dissociate in the heat below and then chemically attack gases found in the lower atmosphere.

The presence of SO_2 and not COS as the major sulphur-bearing compound in the lower atmosphere is a major surprise. Previously it was thought that the abundance of COS indicated that the CO abundance in the atmosphere is buffered by certain minerals at the surface, and its value was representative of the amount below the clouds. These new measurements suggest the CO is generated almost entirely by photochemical processes and no significant amounts of COS should be expected.

Since Venus and the Earth are nearly planetary twins, why are their atmospheres so different? This is not simply an academic question. The two planets must have had several basic similarities in their initial state and the processes that have generated Venus's current inhospitable environment might still occur on Earth.

Venus is much nearer to the Sun and receives twice the amount of incident sunlight that reaches the Earth. This additional heat source would have quickly produced an efficient greenhouse process, increasing the atmospheric CO_2 opacity and raising the surface-temperature to a level which would be too hot for any water-vapour to condense into oceans. Could this happen to the Earth? Man himself is causing the amount of CO_2

to increase in the atmosphere by burning fossil fuel, and, at the same time, reducing the sink for removing CO_2 by world-wide deforestation programmes. He also seems to be devising various methods for depleting the blanket of ozone in the stratosphere. We may shrink from suggesting that the Earth would follow the same evolutionary path as Venus as a consequence. But it is wise to recognize the possibility and do all we can to understand the path followed by each planet throughout its evolutionary history. We still have a great deal to learn about our nearest neighbour for the benefit of mankind on Earth.

Mars

The planet Mars has always interested astronomers, for it is the only member of our solar system that even superficially resembles the Earth. Early telescopes showed distinct markings that could be followed as the planetary surface rotated. The appearance of a disc crowned at the poles with brilliant white caps and light orange and darkish grey–blue coloured areas that change seasonally, provided further information for early astronomers, some of whom speculated that Mars was a hospitable planet and possibly able to support some form of life.

Space missions have vastly improved our knowledge of the planet. We now know that it is a geologically and meterologically (and just conceivably biologically) active planet. The *Viking* lander mission, which is still obtaining valuable information, has made numerous important discoveries. Indeed, we should not measure the success or failure of this mission by the biological investigations (which are still unresolved) but by its contribution to our understanding of planetary processes. There is no doubt that *Viking* has been a resounding success, obtaining fundamental information, amongst other things, on the cause of the Martian channels, the nature of the polar caps, and the size of the ancient atmosphere, all of which collectively help to provide a more complete understanding of the past, present, and possibly the future evolution of the Martian atmosphere.

Atmospheric properties

The atmosphere of Mars is thin compared with the Earth's, with a surface pressure in the neighbourhood of 6 mbar, while a typical terrestrial value is 1013 mbar (1 atm). Surface pressures are primarily a function of altitude and generally higher values are found in depressed areas or basins. A value of 8.9 mbar was determined in the Hellas basin, but pressures as low as 1–2 mbar were derived for the summits of the huge Martian volcanoes (Plate 7).

Temperatures in the thin atmosphere of Mars are much lower than those in Earth. The average surface-temperature is 210 K. It ranges from about 123 K in the polar regions to about 220 K near the equator, and the daytime values reach to almost 300 K for areas of low thermal inertia at midday. The thin atmosphere responds rapidly to the incident solar radiation so that we expect large diurnal temperature variations. These variations, ranging from 187 to 241 K, are indeed much larger than we find on Earth during the course of a day, but possess qualitatively similar features. For example at China Lake, which is part of a broad dry basin in the Mojava Desert, California, the temperature variation ranges from 292 to 311 K. The diurnal phases of temperature maximum and minimum are the same at both the Martian and terrestrial sites. This indicates that convection is the dominant heat transfer mechanism in both cases.

On both Mars and the Earth, the atmosphere consists of gases released from the interior of the planet through volcanism and less violent forms of venting. On Mars the major contribution is thought to have come from the emission of gases by giant volcanoes which lie along the Tharsis ridge. The current Martian atmosphere is mainly composed of carbon dioxide, with traces of water-vapour, carbon monoxide, oxygen, ozone, argon, nitrogen, krypton, and xenon. The abundances of H_2O, O_3, and O_2 are all extremely variable, with both seasonal and geographical variations.

The amount of water-vapour detected in the Martian atmosphere ranges from less than 10^{-4} precipitable centimetres in the high southern latitudes to 10^{-2} precipitable centimetres in the high northern latitudes in midsummer. Since the Earth's atmosphere normally contains two or three precipitable centimetres, this suggests at first sight that the Martian atmosphere is rather dry. If, however, we compare the total amounts of water-vapour above the seven millibar pressure level we find the atmosphere of Mars is very wet. Indeed it is as wet as it can be for the prevailing atmospheric temperatures.

There are important sources and sinks maintaining the observed Martian distribution of water-vapour. In the late summer there is a residual cap at each pole. At the north pole the cap is known to be water ice, whereas the south pole appears to be predominantly composed of dry ice, i.e. CO_2. Sublimation of water ice under the heat of the sun can provide an abundant supply of water-vapour. Although the largest amounts of water-vapour are observed around the northern summer cap, why do we not find similar results at the south pole in summer? It may be due to the dust storms which are currently generated in that hemisphere and seem to transport water-vapour northward.

Water-vapour plays a further major role in the Martian atmosphere, as a key ingredient in its atmospheric chemistry. Unlike the Earth, which is shielded from the harsh ultraviolet rays of the Sun (which are harmful to all forms of life) by a layer of ozone in the stratosphere, the virtual absence of this gas means that the Martian surface is exposed to this energetic radiation. Two important atmospheric chemistry cycles leading to surface oxidation are initiated when CO_2 and H_2O are photodissociated by the ultraviolet radiation:

$$CO_2 + hv \rightarrow CO + O$$
$$O_3 + hv \rightarrow O_2 + O$$

where both O_3 and O contribute to the oxidation of the rocks; and

$$H_2O + hv \rightarrow H + OH$$
$$H + O_2 \rightarrow HO_2$$
$$HO_2 + HO_2 \rightarrow H_2O_2 + O_2$$

where both the superoxide H_2O_2 (hydrogen peroxide) and OH (hydroxyl radical) can also contribute to surface oxidation. The superoxide surface, although a very powerful oxidizing agent, may destroy organic molecules. However this does not in itself preclude the presence of life, which could always reside under the soil or in more sheltered places rather than exposed on the hostile surface of the planet.

Current measurements raise one important unanswered question. The surface pictures provide unequivocal evidence of channels cut by running water; islands at the mouth of an apparent river flow. Were these features really cut by running water, and if so when did it happen and where is the reservoir of water? To answer these fundamental problems we need to know more about the ancient Martian atmosphere.

The noble gases and nitrogen discovered in the Martian atmosphere indicate a much greater amount of atmospheric carbon dioxide at an earlier epoch. Since noble gases are chemically inert and are not removed from the atmosphere through chemical reactions, they provide direct evidence of volatile compounds being released from a planet during its evolution. Chemically reactive substances such as water, carbon dioxide, and, to a lesser extent, nitrogen are less reliable indicators, since they can be trapped in the surface through the formation of nitrates and carbonates.

The best noble gases to use for this purpose are krypton and the two rare isotopes of argon, ^{38}Ar and ^{36}Ar. The more abundant isotope ^{40}Ar is a radioactive decay product of ^{40}K and therefore a better indicator of the amount of potassium than of degassing. However, the *Viking* measurements show that the relative abundances of the two rare Ar isotopes and their ratios to krypton are similar to values measured on the Earth. If this approach is extended to carbon dioxide, then it suggests that the Martian degassing is less complete than on Earth since the current amount of Martian CO_2 is only ten times less than originally present. At the very least this suggests that the ancient atmosphere may have been thicker in the past, with a surface pressure of about 100 mbar.

The nitrogen measurements provide even stronger evidence of a more abundant ancient atmosphere. The *Viking* measurements show an enrichment of ^{15}N over ^{14}N that appears to be due to a differentation of the ligher isotope caused by photochemical processes. The most extreme interpretation of the nitrogen measurements indicates an original carbon dioxide atmosphere as great as the Earth's with a surface pressure of about 1 bar. If all the volatiles had been released at once, water could have covered the planet to a depth of 200 m so that the dust basins would then have been oceans. Thus Mars may be viewed as having outgassed and produced a thick atmosphere early in its history, when water would have run freely over its surface.

Meteorology

The thin Martian atmosphere responds rapidly to radiative and convective processes and to changes in surface temperature. Since the characteristic radiative response time is only two days compared with 100 days on Earth, the large-scale atmospheric motions are strongly controlled by solar heating. However, the tenuous nature of the atmosphere results in inefficient heat transport by winds so that large temperature contrasts exist. There are further complicating factors: the release of latent heat when CO_2 and water-vapour condense to form the polar caps, the frequent dust storms that are created near perihelion affecting the thermal balance and atmospheric stability, and the huge topographic features in the form of deep basins and massive volcanoes.

The Martian weather varies greatly with season and time of day. In winter a massive temperature difference between equator and pole produces brisk westerly winds and creates intense low pressure areas similar to terrestrial systems (Fig. 7.3). In the southern hemisphere Mars and Earth may differ greatly in their weather. The axial tilt of both allows slightly more insolation in the summer polar regions than at the equator at the solstice. Its relative dense and cloudy atmosphere gives Earth a high reflectivity in the polar regions so that less solar radiation reaches the ground at the poles even in midsummer. In the transparent Martian atmosphere a reverse gradient may exist. Light easterly winds will prevail and the usual planetary waves will be absent, so that there will be little in the way of weather. The dominant wind systems will then be tidal due to the atmosphere's rapid response to changes in solar heating.

Since the Martian atmosphere is close to saturation, one would expect clouds to be plentiful. There is nothing resembling the dense, sharp-edged cumulus clouds we know on Earth. The Martian clouds are of four general types: convective clouds, wave clouds, orographic clouds, and fogs. One of the most surprising atmospheric observations made by the *Viking* spacecraft was early morning ground fogs in several low-lying areas, created when ground frost is vapourized by the early morning sun.

But water is not the only substance capable of forming clouds on Mars. In the polar regions in winter, and at high altitudes, the temperatures can fall low enough for carbon dioxide to condense in a dry-ice cloud later. It is also possible that dry ice snowstorms may occur in winter and help create the seasonal dry-ice polar caps. However, much of the dry ice in the caps is probably deposited when carbon dioxide gas comes in direct contact with the cold Martian soil and condenses.

The interaction between the polar caps and the atmosphere has a unique effect upon the meteorology of Mars. About 20 per cent of the carbon dioxide in the Martian atmosphere is cycled between the cap and atmos-

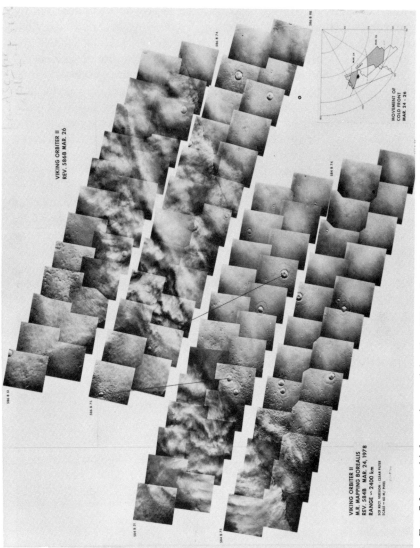

FIG. 7.3 A cold front observed by the *Viking 2* orbiter on revolution 584 in the Borealis region.

phere each season, causing a corresponding variation in atmospheric pressure. The pressure variation is not local but is observed throughout the atmosphere and has been measured at both the *Viking* landing sites.

Without doubt the dust storms which regularly occur at perihelion, and sometimes engulf the entire planet, are the most dramatic aspect of the Martian meteorology. During 1977 at least 36 storms were observed by *Viking*, and two of them developed into global disturbances (Fig. 7.4). What generates the strong winds for such huge storms? It is significant that the Martian orbit is more elliptical than the Earth's. The amount of solar energy impinging on Mars is 40 per cent greater when the planet is closest to the Sun than when it is farthest away. The corresponding figure for the Earth is only 30 per cent. This probably would not account for the huge wind velocities alone, since winds of at least 30 to 60 m s^{-1} are required to raise the dust on Mars, compared with velocities as low as 6 or 7 m s^{-1} in many terrestrial deserts.

The continuous monitoring of the Martian atmosphere by the *Viking* instruments has shown that there are two important mechanisms involved in the generation of these storms, with one more dominant than the other at a particular Martian season. The early major dust storms observed by *Viking* many months before perihelion coincided with the retreat of the south polar cap. At this time a large temperature gradient (of about 80 K) would exist between the newly exposed soil and the cap itself. Near large-scale topographic features, such as the Argyre basin, this would induce a

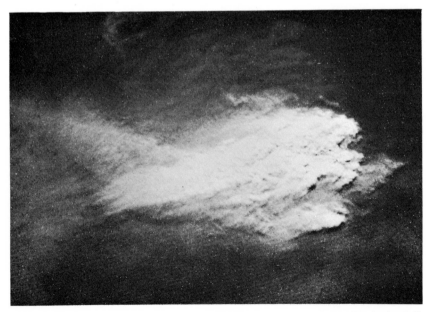

FIG. 7.4 A local dust storm observed by *Viking 2* orbiter on revolution 211 in the Solis Planum region.

wind strong enough to lift the local dust and generate the first storms observed. Indeed during one rotation about four local storms were seen on a ring around the retreating polar cap which then linked to become the first major storm of 1977. Currently, the storms occurring in the northern hemisphere remain local, isolated disturbances. These local storms are also created by temperature gradients induced by passing cloud fronts such as the one seen in Fig. 7.3.

A second important mechanism generating storms at the time of perihelion is related to a special property of the Martian meteorology which occurs during the southern hemisphere summer, namely the dominance of the tidal winds over the planetary scale waves. The strength of the winds is very sensitive to the atmospheric heating, which itself depends upon the amount of dust in the atmosphere. If the atmosphere is sufficiently laden with dust then the winds will grow in intensity until they can raise the dust by themselves. Presumably, this regenerative process begins in localized areas, where the tidal winds and local topographic disturbances give rise to storms or dust devils which are expected to be widespread on this desert-like planet. Certainly we may expect the atmosphere to be constantly laden with dust, a conclusion which is supported by the pictures of a pink sky taken by the *Viking* lander camera.

In spite of these occasional moments of dramatic atmospheric activity, the wind speed at the lander sites has only exceeded the threshold required to lift surface material into the atmosphere on a few occasions. However, dust is redistributed over the seasons, and in Fig. 7.5 we can see changes that have occurred at the Chryse Planitia landing site within two years.

Since a dust-laden atmosphere is warmer than a clear one, dust storms will affect the condensation of water and dry ice at the polar regions. This probably accounts for the layered deposits that are observed, particularly at the south pole, and also for the asymmetry in observed pressure patterns through the Martian seasons.

Climate change

The *Viking* images show examples of features shaped like tear-drops that are unmistakably ancient islands in old stream beds, of water-sculpted terrain and of water-lines along the shores of channels. Their geomorphology would suggest that the channels have been cut by a fluid agent and it is difficult to suggest an alternative to liquid water. Other images show channels that seem to emerge from the head walls of canyons just below regions of collapsed terrain. It is thought that these channels could be caused by slow seepage of underground water or wholesale melting of subsurface ice, which causes the collapse of the region where the water originated.

It is certainly possible that these water-sculpted features resulted from cataclysmic events such as meteorite impacts, or volcanic explosions, since Mars does possess a set of giant volcanoes along the Tharsis ridge. If a large quantity of water was suddenly injected into the atmosphere as a result of meteorites or volcanic activity it might stay there long enough to spread over the planet. Subsequent rain would lead to widespread channel-formation.

Was there a single period of erosion, or could water have been flowing on the planet on a number of separate occasions? We have seen that the ancient atmosphere was once thick enough for large quantities of water to flow on the surface. Could it happen again?

FIG. 7.5 Observations at the Chryse Planitia landing-site showing changes in the surface material. The top picture was taken on Sol 25 on 15 August 1976, and the lower one on sol 771 on 20 September 1978. The change A appears as a small circle-like formation on the side of a drift in the lee of Whale Rock. A slump (B) occurred between sols 74 and 183.

It is possible that the surface layers act as a reservoir for volatile materials. We know that water-ice is locked up in the polar regions and some carbon dioxide may be contained within the surface layers also. How is it possible to release these volatiles into the atmosphere? There are three important cyclic changes in a planet's movement which produce a variation in the amount and distribution of solar radiation incident upon the planetary disc, but which do not affect the integrated flux of heat received by the whole planet in the course of a year. They are changes in: the eccentricity of the orbit, the obliquity of the axis of rotation, and the precession of the longitude of perihelion. The variations are huge for Mars compared with the Earth. For example, the Martian obliquity (the tilt of the rotation axis with respect to the planet's orbital plane) varies by $\pm 10°$ over 1.2×10^6 years, while the corresponding changes for the Earth are only $\pm 1°$ over 4×10^4 years. Recent studies have shown that astronomical perturbations to the Earth's rotation play a critical role in initiating the advance and retreat of the terrestrial ice ages. During the extremes of the Martian obliquity the annual insolation at the polar regions may vary by more than 100 per cent, which must have important climatic implications. If there are reservoirs of carbon dioxide available in the surface then at some future time the additional radiation striking the surface will release it and create a more massive Martian atmosphere. The climate will simultaneously move away from its current ice-age conditions.

There is no doubt that the climate of Mars is potentially highly variable. The changes in obliquity will affect the redistribution of H_2O and the generation of dust storms, which at some epochs will be initiated in the northern hemisphere rather than as is the case now, in the south. The possibility of future wet episodes and a hospitable environment on the planet are exciting speculations. However, there is little doubt that understanding the mechanisms that control and influence the Martian climate will not only produce valuable information in its own right, but will also help our understanding of all planetary atmospheres as we strive to understand the reasons for climate change on the Earth.

Phobos and Deimos

Besides enriching our knowledge of Mars itself, the current *Viking* mission has provided a wealth of information on the two tiny satellites, Phobos and Deimos.

Several close encounters by the spacecraft with the satellites have revealed some startling features on their surfaces (Fig. 7.6). The *Viking 1* orbiter passed less than 100 km from Phobos in February 1977, and 300 km in May. In October 1977, *Viking 2* orbiter flew to within 23 km of Deimos. When one considers that these ellipsoidal-shaped objects have dimensions of 12 km (Deimos) and 27 km (Phobos) along their longest axis, the precision of these passes is amazing.

The measurements suggest that Phobos is made of dark low-density material (about $2\,\mathrm{g\,cm}^{-3}$) similar to Type II carbonaceous chondrites (a certain type of meteorite). This suggests that this moon did not form in orbit around Mars, but instead is a fragment of an asteroid originating from deeper interplanetary space and captured while passing close to the planet.

Phobos' surface has strange grooves, typically 100–200 m wide and 20–30 m deep. These grooves seem to be associated with the large crater Stickney, which is about 10 km in diameter. At highest resolution the grooves appear not as simple cracks but as lines of more or less contiguous pits. The possible explanation of this appearance is the sifting of loose rocks into deep cracks, while another theory suggests that the loose material has been ejected by gas escaping through vents along the cracks.

Some of these features may be related to the large impact that formed Stickney. Are these grooves common on small solar-system objects? Perhaps not, since they are absent on Deimos. Why do large amounts of loose material litter Deimos but not Phobos? One possibility is that the surfaces of the two satellites have significantly different mechanical pro-

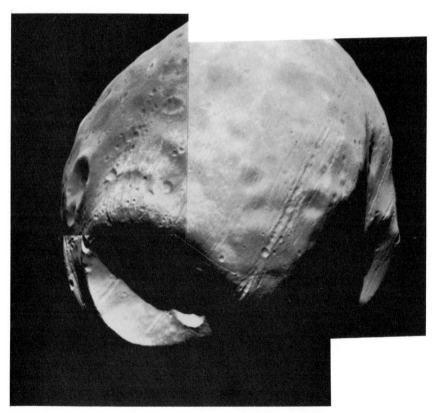

FIG. 7.6 The tiny moon Phobos as observed by Viking *Orbiter 1* on revolution 854. The photo mosaic shows the front side of Phobos that always faces Mars. Stickney, the largest crater on the body is at the left near the morning terminator.

perties. For example, an impact might produce a larger fraction of low-velocity fragments on Deimos than it would on Phobos. The latter is so close to Mars that it may be easier to remove material from its surface than from the more distant satellite Deimos. Phobos is actually within the Roche limit of Mars, and remains intact because it is held together by cohesive forces as well as self-gravity.

There is little doubt that even these tiny bodies orbiting Mars have posed many unresolved questions. The spacecraft exploration of Mars and its environment has been rewarding for all scientists, but there is still a great deal more to be done by future missions.

Jupiter

Any theory of the origin of the solar system must explain the extreme differences between the terrestrial planets (Mercury, Venus, Earth, and Mars) and the major planets beyond the asteroid belt, Jupiter, Saturn, Uranus, and Neptune. These planets are huge, rapidly rotating, low-density objects with optically thick reducing atmospheres, totally covered with cloud. The low density suggests that, like stars, they are entirely composed of light elements: hydrogen, helium (thought to be the principal components of the initial solar nebula), carbon, and nitrogen. This contrasts with the iron, nickel, and silicates which constitute the cores of the inner planets. Furthermore, the four large Galilean satellites have densities which decrease with distance from Jupiter so that they resemble a mini solar system. Consequently, these outer planets, and Jupiter in particular, may hold the key to the formation of the solar system.

The exploration of the outer planets began with the *Pioneer 10* and *Pioneer 11* flybys of Jupiter in December 1973 and 1974. Their brief accomplishments have been dramatically overshadowed by the startling discoveries made by the *Voyager* spacecraft. There is no doubt that our understanding of Jupiter and its satellites has increased beyond the realms of human imagination during the first few months of 1979.

Magnetosphere and environment

The first detection of a Jovian magnetic field was made by Earth-based radio astronomical measurements in the 1950s, when radio frequency emission from the planet was discovered. Emission is confined to two relatively broad regions of the spectrum at decimetric (tenths of metres) and decametric (tens of metres) wavelengths. A major component of the decimetric emission is thermal radiation emitted by the entire planetary disc while the decametric radiation is non-thermal and is produced by the magnetic field of Jupiter. It consists of synchrotron radiation emitted by electrons moving near the speed of light in the Jovian magnetic field. The relativistic electrons follow helical paths, spiralling along the magnetic lines of force, and radiating away part of their energy as they travel between the magnetic poles.

Table 7.2
Physical properties of the planets

	Mercury	Venus	Earth	Mars	Jupiter
Mean distance from Sun (millions of kilometres)	57.9	108.2	149.6	227.9	778.3
Mean distance from Sun (astronomical units)	0.387	0.723	1	1.524	5.203
Period of revolution	88 days	224.7 days	365.26 days	687 days	11.86 years
Rotation period	59 days	− 243 days Retrograde	23 hours 56 minutes 4 seconds	24 hours 37 minutes 23 seconds	9 hours 50 minutes 30 seconds
Orbital velocity (kilometres per second)	47.9	35	29.8	24.1	13.1
Inclination of axis	<28°	3°	23°27′	23°59′	3°05′
Inclination of orbit to ecliptic	7°	3.4°	0°	1.9°	1.3°
Eccentricity of orbit	0.206	0.007	0.017	0.093	0.048
Equatorial diameter (kilometres)	4880	12 104	12 756	6787	142 800
Mass (Earth=1)	0.055	0.815	1	0.108	317.9
Volume (Earth=1)	0.06	0.88	1	0.15	1316
Density (Water=1)	5.4	5.2	5.5	3.9	1.3
Oblateness	0	0.0004	0.003	0.009	0.06
Atmosphere (main components)	None	Carbon dioxide	Nitrogen Oxygen	Carbon dioxide	Hydrogen Helium
Mean temperature at visible surface (degrees K) S=solid, C=clouds	700 (S) day 100 (S) night	240 (C) 737 (S)	295 (S)	210 (S)	123 (C)
Atmospheric pressure at surface (millibars)	10^{-9}	90 000	1000	6	?
Surface gravity (Earth=1)	0.37	0.88	1	0.38	2.64
Known satellites	0	0	1	2	14

The decametric radiation is intermittent, and its intensity is modulated by the Galilean satellite Io. Except near the planet the major component of the field is dipolar, like the Earth's. However, its direction is opposite, so that a terrestrial compass on Jupiter would point south. The axis of the dipole is inclined to the rotational axis by about 10.8°, and the centre is displaced from the centre of the planet by about 0.1 of Jupiter's radius, R_J. At cloud-top level the field strength varies from 3 to 14 Gauss, compared with the Earth's surface field strength of 0.3 to 0.8 Gauss. Within 3 R_J, the field is strongly non-dipole, with quadrupole and octopole components dominating.

The magnetic field, with its entrained plasma, makes up the Jovian magnetosphere. This is huge, highly variable in size, and lies beyond the complex ionosphere. It can extend to as much as 100 R_J. Indeed, if it were seen from Earth with the naked eye, the Jovian magnetosphere would appear as large as the Sun. Despite its size, its geometry is similar to the Earth's. The pressure due to the solar wind diminishes as the square of the distance from the Sun. This is an important factor contributing to the enormous size of the Jovian magnetosphere, since the solar wind is 25 times weaker in this part of the solar system than at the Earth. Relatively minor changes in the pressure of the solar wind can cause large variations in the position of the magnetopause by more than a factor of two in radius. For the Earth's magnetosphere to shrink or expand by such a large factor is exceedingly rare and would be expected only during the most intense magnetic storms.

On Jupiter the electromagnetic environment has a further distinguishing feature, namely that the four Galilean satellites, Amalthea, and the newly discovered ring (see below), all reside in this hostile environment of charged particles. This is quite different from the Earth, where our Moon resides in the magnetotail. As a consequence, the interaction of charged particles with Jupiter's inner moons produces features that are unique to the Jovian environment.

The recent *Voyager* observations have shown that auroral activity is far more intense on Jupiter than previously expected. It dominates the solar absorption processes at ultraviolet wavelengths on the sunlit hemisphere. On the dark side of the planet an auroral arc 30 000 km in length has been observed, which is the largest yet seen on any object in the solar system.

Atmospheric properties

The visible appearance of the planet is a banded structure of alternating light and dark bands running parallel to the equator. The colours of the bands change rapidly. The Great Red Spot is a predominant feature in the planet's southern hemisphere, Plate 8.

The atmosphere of Jupiter is mainly composed of hydrogen and helium in approximate solar abundances with small but important amounts of strange and exotic constituents such as ammonia, methane, water-vapour, ethane, acetylene, phosphine, germanium tetrahydride, and carbon monoxide. Most of these constituents have been found in and around the cloud tops, and it is thought that these hydrocarbon species (and others still to be found), may play an important role in the colourful appearance of the planet by a process of complex atmospheric chemistry.

Since ammonia is a constituent of the Jovian atmosphere it will condense into a layer of cloud at some tropospheric level. Indeed we believe that the ammonia clouds mark the top of the bright zones, and that these clouds are absent at the higher levels of the dark belts. Beneath the ammonia clouds we expect other cloud-layers to form, involving NH_4SH, H_2O ice, and NH_3 solution. Determining the precise composition of the belts must await analysis by the space-probe mission in the mid 1980s.

As well as clouds, aerosols are formed from photochemical reactions in the upper troposphere and stratosphere. The observations suggest that the lightning storms take place on a planet-wide scale. Indeed ultraviolet radiation and lightning may be important energy sources in the chemical reactions that take place in the planet's atmosphere. It is possible that the observed ethane and acetylene are produced by ultraviolet photolysis of methane. While at high levels most of the H_2O is probably removed by condensation with NH_3 to form a cloud of NH_4SH, at levels below 90 km it may be dissociated by ultraviolet light and lightning. This could initiate the following reactions leading to the production of elemental sulphur

$$H_2S + hv \rightarrow HS + H$$
$$HS + HS \rightarrow H_2S + S$$
$$\text{or} \qquad H_2 + S_2.$$

Since hydrogen polysulphide H_xS_y or ammonium polysulphide $(NH_4)_xS_y$ are generally yellow, orange, or brown according to the local temperatures, they may account for some of the Jovian colours. A key piece of evidence to support this suggestion would be the identification of H_2S. We are still searching for this with spectrometers aboard *Voyager* spacecraft.

Meteorology

The weather systems on Jupiter are very different from those found on the other planets, and a completely new perspective has been given by the incredible high resolution images obtained during the *Voyager* encounters. The planet was observed continuously by *Voyager 1* from 6 January until encounter on the fifth of March, with resolution increasing dramatically with every passing moment until, at the time of closest approach, scales of 3 km could be observed. These observations have coupled together space and time, so that a complete four-dimensional picture of the planet's atmospheric motions has been obtained. We are now in a position to unravel Jupiter's complicated meteorology.

Jupiter emits nearly twice the energy it receives from the Sun and so is rather more like a failed star with its own internal heat source than a typical planet. This additional energy source has an important role in Jovian meteorology since it means that the planet's weather is affected both by the internal heat source and the external solar radiation. Also the planet's rotation has a strong influence, since the rotation period is approximately 10 hours. Consequently, on a large scale, the atmospheric motions are predominantly parallel to the equator, i.e. zonal, but, as can be seen from Plate 8, there are regions where significant departures from this conclusion exist. However, on a global scale, there is little, if any, pole-to-equator gas flow, or energy transfer, at visible cloud levels. Indeed the *Pioneer 11* measurements showed there was only a very small difference (of no more than 3 K) between the effective temperature of the planet at the equator and at the poles. The imaging system observed a gradual breakdown of the dominant symmetric belt and zone pattern beyond $\pm 45°$ latitudes, where internal heating becomes the dominant driving mechanism. None the less

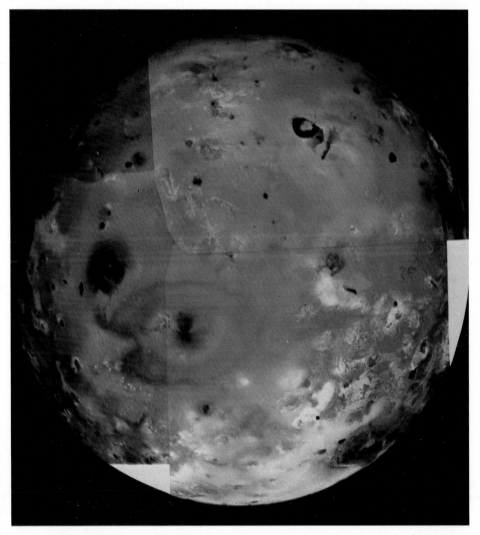

X. A global mosaic of Io obtained by *Voyager 1* on 4 March 1979. The colours are exaggerated to emphasise the surface contrasts.

XI. Europa observed by *Voyager 1* on 4 March 1979, showing the face centred at a longitude of approximately 300°.

XII. Ganymede observed by *Voyager 1* on 4 March 1979.

the flow still seems reasonably well organized and approximately zonal there.

It is important to note the asymmetry in the cloud morphologies between the northern and southern hemispheres. Analysis of the *Voyager* data confirm the Earth-based observations of the positions of the jets between the belts and zones. These jets are symmetric in their strength about the equator. This suggests that deeper motions at depths of hundreds of kilometres are important in maintaining the jets and long-lived features.

The temperature contrasts between belts and zones is small. However, the location of maximum contrast does vary, and between the *Pioneer 10* and *Pioneer 11* flybys of 1973 and 1974 and the *Voyager* flyby of 1979, it shifted hemispheres. Previously the largest contrast (of about 3 K) occurred between the South Equator Belt and the South Tropical Zone, while now it is at equivalent latitudes in the northern hemisphere. These temperature contrasts may be due to the release of latent heat of condensation in regions where the atmospheric gases are rising. In the past scientists concluded that the zones are regions of anticyclonic vorticity and high pressure, and the belts are low pressure systems. These zones and belts produce alternate bands of easterly and westerly winds, concentrated into jets at their interfaces. We find that the highest speed jet lies in the North Temperate Current, seen as a thin brown line in the broad zone in Plate 8. Here westerly wind speeds are typically about 150 m s^{-1}. At the equator, zonal wind speeds are about 110 m s^{-1}, while the train of spots east of the Great Red Spot have been seen to approach the Spot from the east at speeds of about 55 m s^{-1}.

The overall picture of Jupiter shows a great variety of cloud morphologies of varying sizes, colours, and shapes, of instabilities in the form of vortices, and of course, the Great Red Spot. The regularity in the spacing of some of the cloud spots suggests some form of wave interaction. A remarkable feature of Jovian cloud patterns is their longevity, in spite of the apparently turbulent flows in which they are found. On Earth, a cloud feature rarely persists for more than a few days unless it is tied to a topographic feature. But Jupiter is a gaseous fluid throughout, so there can be no such feature below the visible cloud level. The long lifetime may be due to the slow dissipation rate which is equal to a cooling rate of 10 K per year compared with 1 K per day on the Earth. Together with a radiation relaxation time of 10 years, this implies that temperature anomalies are likely to take a long time to disappear by radiation processes. Furthermore, friction appears only in the form of viscosity, which is a much weaker force than that of direct interaction with a solid surface.

Understanding the Great Red Spot has occupied a great deal of time since it was discovered in 1664 by Hooke and Cassini. *Voyager* has proved true the statements that I have been making for many years, namely, it is not unique as a meteorological feature. Indeed the observations (Plates 8 and 9) show that morphologically it has the same characteristics as the white ovals and many of the other spots observed. The Great Red Spot and these other white spots are all high-pressure areas, and associated with each of them is a disturbed region on the west. It appears that the features act as a barrier to the flow approaching from the west, and an obstacle to the flow approaching from the east, so that a wake is created between strong shears. This situation is clearly shown in Plate 9. It seems that this is a type of Jovian 'blocking anticyclone', a process which in the terrestrial atmosphere has been important in the recent fluctuations in our own weather system (such as the drought of 1976).

The high-resolution images show that there is now a general circulation around the Great Red Spot with a period of about six days, compared with a ten-day period ten years ago. The Great Red Spot has also shrunk by 30 per cent in the last ten years and is now only 21 000 km long and 11 000 km wide. The vorticity is about 2.5×10^{-5} s^{-1}. The *Voyager* studies of the great red-spot, white ovals, and smaller white cloud spots show that there is a divergent flow in their interiors and an upward propagation of energy from each of these features which decreases in magnitude with the size of the cloud.

The colour of the Great Red Spot remains puzzling. This feature is unique, with the exception of small red spots occasionally to be seen in the Northern Tropical Zone. The red colour may be related to the presence of phosphine in the atmosphere. The Great Red Spot is known to be elevated relative to its surroundings, so that if material is brought up above the ammonia cloud level the action of ultraviolet light may produce red phosphorous by the reactions:

$$PH_3 + hv \rightarrow PH_2 + H$$
$$PH_2 + PH_2 \rightarrow PH + PH_3$$
$$PH + PH \rightarrow P_2 + H_2$$
$$P_2 + P_2 + M \rightarrow P_4(g) + M$$
$$P_4(g) \rightarrow P_4(s)$$
$$H + PH_3 \rightarrow H_2 + PH_2$$

However, C_2H_2 and C_2H_4 could act as scavengers to the phosphorous reactions and reduce the amount produced. Consequently, red phosphorous may be produced only in those locations where C_2H_2, and C_2H_4 have low concentrations. This may help to account for the fact that only localized regions appear with this red colour. But why is there not a multitude of red spots? It may simply be that small features do not penetrate sufficiently deep to where phospine is available.

The dynamics of the Jovian atmosphere are of interest not only to meteorologists but to exobiologists and chemists also, for they have long debated whether life or its chemical precursors may be found on the giant planet. Some believe that organic molecules may account for the observed colours, although, as we have already indicated, inorganic sulphur compounds can also provide an adequate explanation. The presence of lightning storms all over the planet adds a further dimension to these speculations. To settle this issue we need to know the vertical velocity structure, and the convective state of the Jovian atmosphere.

Ring properties

Without doubt, one of the most amazing discoveries found in the *Voyager* images is the ring of material that surrounds Jupiter. Jupiter is the third planet in the outer solar system discovered to possess a ring.

The Jovian ring is sparsely populated with disc material and is estimated to be less than 30 kilometres thick, 6500 kilometres wide, and extending more than 140 000 kilometres above the cloud tops. Minute particles circle in the ring and extend down to the cloud top (Fig. 7.7). It has a stellar magnitude of approximately +22. This ring lies within the Roche limit for Jupiter. However the origin and stability of the ring pose a fundamental problem. The amount of material there is very small and unlikely to be the remains of a satellite. These rings have more in common with the Uranus rings than with those around Saturn. There is also the exciting possibility that the Jovian rings are a transitory feature. *Voyager* observations make more natural the model of Jupiter as a mini solar system. Surrounded by an array of satellites, it now seems to have its own ring system.

FIG. 7.7 An image of the dark side of Jupiter taken by *Voyager 2* on 10 July 1979, showing the ring and bright halo around the planet. The ring is unusually bright owing to forward scattering of sunlight by small particles.

Inner satellites

Although the satellite, Amalthea, was discovered in 1892 by Barnard, it has never been seen as anything but a tiny point source until the *Voyager 1* flyby (Fig. 7.8). We now find this fragment of material is not spherical but oblate: about 130 km by 170 km, and extremely dark, with an albedo of less than 10 per cent. At a resolution of about 8 km it is possible to resolve craters on its surface. Indeed, the irregular shape of this satellite may reflect a long history of impacts. An important question yet to be answered is whether its red colour is characteristic of the bulk of Amalthea, or whether it results from interactions with the magnetosphere that modify the surface properties. The satellite maintains its long axis pointed towards Jupiter as it rotates about the planet every 12 hours.

In 1610 Galileo Galilei turned his primitive telescope toward Jupiter and made the discovery that the giant planet was surrounded by four moons. This unexpected observation startled the world at that time. However, their feeling of surprise has been far exceeded by our own at the first close-up observations of these satellites by *Voyager* in March 1979. In a matter of a few days they changed from being point sources of light to planet-sized objects with surface features resolved to less than 1 km.

FIG. 7.8 The tiny satellite Amalthea taken by *Voyager 1* on 4 March 1979.

Since the discovery that it modulated the Jovian decametric radiation Io has always been an enigma. *Pioneer 10* showed that this strange satellite possessed a thin atmosphere with a surface pressure of 10^{-8} bars as well as its own ionosphere. At approximately the same time, ground-based observers discovered that Io was surrounded by a cloud of sodium atoms sputtered from the surface by the high-energy particles that collide with the body. Later observations have found magnesium and potassium in this cloud. The cloud has a vertical extent of about $2\,R_J$.

Table 7.3

Physical properties of the inner Jovian satellites

Satellite	Mean opposition Mag	Radius (km)	Density $(g\ cm^{-3})$	Mass $(10^{23}\ g)$
Io	5	~ 1820	3.52	891
Europa	5.3	~ 1500	3.45	487
Ganymede	4.6	~ 2635	1.95	1490
Callisto	5.6	$\sim 2500 \pm 100$	1.62	1065
Amalthea	13	$(\sim 130 \times 170$ km)	?	?

The observations of Io show it to have an unexpectedly smooth surface, but with a considerable amount of high relief estimated to reach to an altitude of 10 km (Plate 10). The highly coloured nature of the surface is thought to consist of mixtures of salts and sulphur materials.

The explanation of these strange features comes from the discovery by *Voyager 1* in March 1979 that Io has at least eight active volcanoes (Fig. 7.9). During the *Voyager 2* encounter in July 1979 seven of these volcanoes were observed again and six remained active. Solid material appears to be thrown out of these volcanoes to an altitude of more than 200 km in an enormous fountain with an ejection velocity from the volcanic vent of about 2500 km per hour. The material reaches the crest of the fountain in several minutes. It appears that Io's volcanoes are considerably more violent than any known terrestrial volcano.

But why should Io possess active volcanoes? Io is the innermost Galilean satellite. As it moves in its orbit around Jupiter, the orbital eccentricity is modulated by the gravitational forces exerted on Io by the other Galilean satellites. Although still modest by most standards, the forced eccentricites coupled with the enormous tides induced by Jupiter lead to magnitudes of tidal dissipation that dominate the thermal history of Io. Heating in the solid mantle of the satellite melts material near the liquid core, thereby diminishing the mantle thickness. As a result the thinner elastic shell undergoes further deformation and produces further heating. The result of this runaway melting process is a planet with a large molten core and a solid outer shell, the thickness of which is limited by conduction of the internally generated heat to the surface, or possibly by fracturing and the onset of inelastic behaviour. The high internal temperatures imply a molten core with a radius of perhaps one-third that of the satellite. This suggests that Io is probably the most intensely heated terrestrial body in the solar system.

There are no such tidal forces acting on the Earth's moon which, although approximately the same size and mass as Io, is a cold, dead world. Today the Moon is solely heated internally through the decay of radioactive isotopes. This is insufficient to produce a large molten core and volcanic activity.

The observations of volcanic activity on Io immediately accounts for the smooth surface appearance. Ejected volcanic material has covered up any meteor impact craters that have occurred. In addition to the active volcanoes themselves, lava lakes have been discovered as local 'hot spots' at temperatures of about 350 K, compared with the surrounding terrain at about 100 K.

It is likely that some of the material ejected from these volcanoes forms a torus in the Io orbit. A cloud of doubly ionized sulphur was detected during the *Voyager* flyby. Since this material is ionized it is tied to, and therefore

FIG. 7.9 An active volcano on Io observed by *Voyager 1* on 4 March 1979.

rotates with, the Jovian magnetic field. It therefore orbits Jupiter four times faster than Io itself, at the rotation rate of Jupiter.

There is little doubt that Io is a strange world. Since it has a molten core there is every likelihood that it also has a magnetic field. For this to be detected, a spacecraft will have to fly closer to Io than *Voyager* managed to do, but that is something for the future.

Europa is a startling satellite and is the smoothest body ever seen in the solar system. The *Voyager* picture (Plate 11 and Fig. 7.10) show no evidence of topography such as mountains and volcanoes. The satellite is covered with a network of cracks that intersect regularly over the surface. Europa is covered by a layer of slushy ice, probably only about 100 kilometres thick, which covers up any craters that may form, thus explaining their apparent absence. The state of this surface is due to tidal heating, as on Io, but at least a factor of ten times less efficient. Some of the surface lines have a white interior which may therefore be new ice, forcing its way through the fissures. We have never seen a planetary body like this before.

The two outer satellites Ganymede and Callisto are quite different from the inner pair. They are much lower density objects and are thought to be predominantly icy bodies. The images show dirty water-ice surfaces riddled with craters, reflecting the heavy impact of meteors during past epochs.

On Ganymede (Plate 12) the crater density is less than that on Callisto, but it possesses spectacular ray features, hundreds of kilometres wide. There are many bright spots, possibly regions of fresh ice uncovered by recent impacts. A characteristic of the Ganymede surface is the peculiar system of sinuous ridges and grooves (Fig. 7.11), which are probably due to deformations in the ice. They appear to cover the entire planet, and may therefore be caused by tectonic activity.

Callisto (Fig. 7.12) shows the most densely cratered surface in the solar system. The craters do not seem to have the same sharp rims and deep floors as the craters on Mercury and the Moon. Instead they appear quite shallow. The fact that neither Callisto nor Ganymede show any topographic relief suggests their crusts are not rigid enough to support such features. On Callisto, the regularity of the size of the observed craters is directly related to the properties of the crustal material.

The huge basin observed on the surface of Callisto (Fig. 7.12) is about 600 km in diameter. The numerous concentric rings that surround the basin probably result from the response of the icy crust to the shock wave created by the large impact that formed the basin itself. These rings extend outward to distances of more than 1000 km. The central portion of the basin has become filled in with material through the passage of time and cratered by smaller impacts. The observations by *Voyager* of fireballs entering the Jovian atmosphere are important since they will help provide estimates of the flux of material impacting these satellites.

As our observations of these Jovian moons continue there is little doubt we will confirm *Voyager's* discoveries about these four totally new worlds;

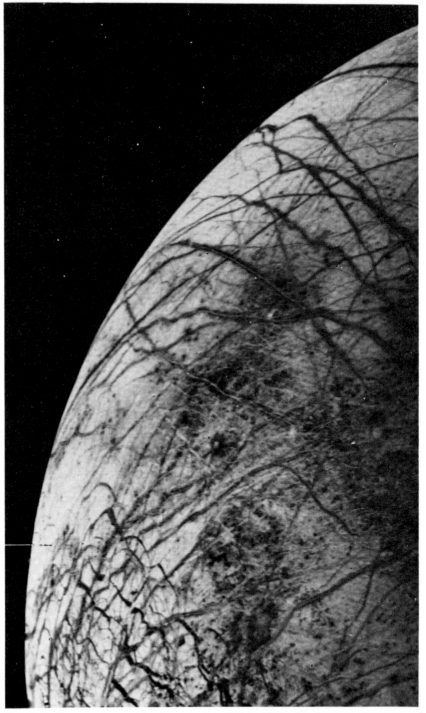
FIG. 7.10 Europa observed by *Voyager 2* on 9 July 1979, showing complex array of streaks indicating that the crust has been fractured and filled by material from the interior.

discoveries which must go far beyond the dreams and speculations of Galileo when he made the startling discovery of their existence.

Exciting planetary prospects of the future

A dramatic leap forward in our knowledge and understanding of planetary process can be expected in the coming years. The second *Voyager* spacecraft reached Jupiter in July 1979. In 1980 *Voyager 1* reaches Saturn and will conduct the first detailed look at Titan, while in 1981 the second spacecraft will conduct its complementary study of the Saturnian system. There is every hope that it will continue to Uranus in 1986 and Neptune in 1989. *Pioneer* is still in orbit studying our nearest neighbour Venus, while *Viking*

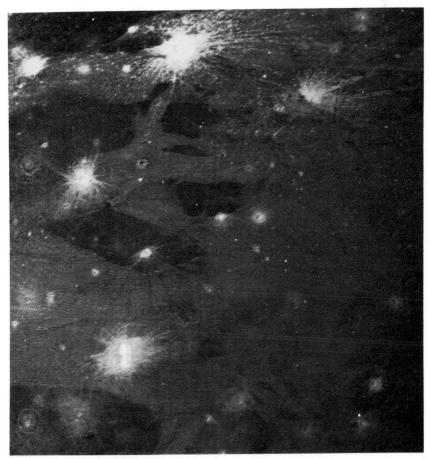

FIG. 7.11 A high resolution image of a segment of Ganymede observed by *Voyager 1* on 5 March 1979. The image shows detail with a resolution of 4.5 km.

is concluding its orbital mission to Mars. It is possible that the instruments at the Chryse landing-site may continue in operation for another ten years. Still farther ahead there are missions to Venus to map the surface, cometry flybys, Jupiter orbiters, and the prospect of studying the meteorology of the planets with the Space Telescope.

The next decade is destined to provide an unparalleled advancement in our knowledge and understanding of planetary processes. The picture emerging is one full of surprises pointing to a vigorous, continuing evolutionary process. We are living in a golden age for the exploration of the solar system.

Acknowledgements

The pictures used here have been provided by NASA. This research is supported by the Science Research Council.

FIG. 7.12 A mosaic of Callisto obtained from images taken by *Voyager 1* on 6 March 1979. The large impact basin has a diameter of 600 km.

Bibliography

It is not possible to list all the papers that are related to the subject discussed in this article. Consequently I have been very selective in the individual articles referenced. However, the specific journal volumes and books that contain numerous articles on the results from recent space missions are clearly indicated.

MERCURY

J. geophys. Res. **80**, 17 (1975).
MURRAY, B. (1975). *Scientific American* **233**, 58–69.
Science, N.Y. **195**, 141–80 (1974).

VENUS

AVDUEVSKY, V. S., MAROV, M. Ya., MOSHKIN, B. F., and EKONOMOV, A. P. (1973). *Venera* 8: measurements of solar illumination through the atmosphere of Venus. *J. atmos. Sci.* **30**, 1215–18.
—— ——, and ROZHDESTVENSKY, M. K. (1970). A tentative model of the Venus atmosphere based upon measurements of *Veneras 5* and *6*. *J. atmos. Sci.* **27**, 561–79.
—— —— ——, BORODIN, N. F., and KERZHANOVITCH, V. V. (1971). Soft landing of *Venera 7* on the surface of Venus and preliminary investigations of the Venus atmosphere. *J. atmos. Sci.* **28**, 263–4.
MAROV, M. Ya. (1973. Preliminary results on the Venus atmosphere from *Venera 8* descent module. *Icarus* **20**, 407–21.
—— (1973). *Venera 8*: measurements of temperature, pressure and wind velocity on the illuminated side of Venus. *J. atmos. Sci.* **30**, 1210–14.
VAKHNIN, V. M. (1968). A review of the *Venera 4* flight and its scientific programme. *J. atmos. Sci.* **25**, 533–4.

Mariner 10 flyby

Science, N.Y. **183**, 1289–1321 (1974).

Venera 9 and *Venera 10*

Cosmic Res. **14**, 573–701 (1976).
KELDYSH, M. V. (1977). Venus exploration with the *Venera 9* and *10* spacecraft. *Icarus* **30**, 605–25.

Pioneer Venus

Science, N.Y. **203**, 743–808 (1979).
Space Science Rev. **20**, 249–325 (1977).

MARS

Mariner 9

Icarus **17**, 289–327 (1972).
Science **175**, 293–323 (1972).

Viking

J. Geophys. Res. **82**, 3959–4681 (1979).
Science, N.Y. **193**, 759–815 (1976).
Science, N.Y. **194**, 57–109 (1976).
Science, N.Y. **194**, 1274–1353 (1976).

JUPITER

GEHRELS, T. (1976). *Jupiter*. University of Arizona Press.

Pioneer 10 and *Pioneer 11*

Science, N.Y. **183**, 301–24 (1974).
Science, N.Y. **188**, 445–77 (1976).

Voyager

Nature, Lond. **280**, 725–804 (1979).
Science, N.Y. **204**, 945–1008 (1979).
Space Sci. Rev. **21**, 75–376 (1977).

8

New ways of seeing the Universe

F. G. SMITH

My subject is the spectacular extension of observational techniques which we have seen in astronomy over the last 25 years. These techniques now extend through practically the whole electromagnetic spectrum involving radio telescopes, radio interferometers with intercontinental baselines, balloon-borne gamma-ray telescopes, and X-ray telescopes on satellites. Spectacular indeed, and apparently overwhelming the old-fashioned optical astronomy, confined as it is to the visible and near-visible parts of the spectrum. I shall describe the relation between the old and the new astronomies, and show that the new has not led to the abandonment of the old, but rather to its renewal. The more we look through the newly opened windows on to the Universe, the more we realize the importance of the familiar optical window.

Consider the human eye. It selects less than one octave of the spectrum: by way of contrast the Crab pulsar produces detectable pulses over 49 octaves. What use is it to pick out only one octave? There are in fact two good reasons why the optical range is so useful, and they are so strong that they guided the evolution of vision in the living world. The first concerns information content, and the second concerns technique. I shall dwell on these, for they have also guided the rather more rapid evolution of the new astronomies.

First, consider the availability of information. It happens that our atmosphere is transparent to just those wavelengths of radiation which are most efficiently produced by the Sun. Actually we lose the ultraviolet end of the spectrum—I will have more to say about that—but roughly speaking the spectrum of light from an average star like the Sun peaks very conveniently in the visible range, giving us a means of seeing both terrestrial objects and most types of star in the sky. Also (and this is most important) spectral lines characteristic of many common atomic species lie in this visible range. Emission lines from sodium and mercury, for example, make useful street lamps. We can analyse the light from a star and see spectral lines which provide detailed information on the physical conditions and chemical abundances in stellar atmospheres. By a fortunate coincidence, the radiation from a star at 6000 K, the transparency of the atmosphere, and the common spectral lines, all occur in that one narrow band of wavelengths in the optical region.

What of the technique that allows the eye to detect light? It is more efficient to detect photons than merely to detect the flow of energy, and it turns out that living tissue can adapt itself to become a quantum chemical detector at optical wavelengths, but not so easily outside the optical waveband. Light-sensitive chemicals, such as the rhodopsin of the human eye, are sensitive because suitable quantum levels exist for the absorption of visible photons. Again, refracting materials suitable for lenses are available for living creatures to adopt—watery materials for human eyes, calcite for the trilobites.

The human eye is quite good in another essential quality, angular resolution, in which it achieves about 3 arc minutes; this is possible because the wavelength of light is small compared with the diameter of the pupil, say 500 nanometres wavelength compared with 2 millimetres diameter, giving a few thousand wavelengths across the aperture. This resolution is acceptable enough in everyday life, but not, of course, good enough for most of astronomy, where resolutions now reach down to one thousandth part of an arc second—one milliarcsecond. One milliarcsecond needs an aperture of 100 million wavelengths.

Let us consider the first new window to be opened up—radioastronomy. The usefulness of radio waves to astronomy was not at first obvious, even though Jansky showed in 1932 that there was strong cosmic radio emission, which he was able to associate with the Milky Way. As the techniques for analysing the cosmic waves developed, so their usefulness became more apparent, and techniques developed further. We now have radio-telescopes covering wavelengths from 15 metres right down to below 1 centimetre. These wavelengths are typically a million times longer than those of visible light, and the quantum energy is correspondingly smaller. Most people are familiar with the spectacular techniques which developed in response to the need to collect radiation over a large area, such as the 250 foot Mk I A radio-telescope at Jodrell Bank, and other large reflector radio-telescopes. With large collecting areas and very sensitive radio receivers radio signals can be detected from our own Galaxy, from objects within it such as the pulsars, from other galaxies like our own, and from extragalactic objects such as the quasars.

In an attempt to follow these successes, radioastronomy has undergone further evolution. It was a triumph to discover the quasars, known to be very distant and very condensed energetic sources of radio waves, but for many years the challenge remained—how to draw a high-resolution map of a quasar. The Mark I radio-telescope works at 6 cm wavelength, where it is

1300 wavelengths across, giving a resolution much worse than that of the eye. Ryle at Cambridge developed techniques of interferometry and aperture synthesis which brought the resolution to 1 arc second, revealing a wealth of detail, such as that shown in the radio-picture of the supernova remains Cassiopeia A (Fig. 8.1), whose whole diameter is only 3 arc minutes (less than the resolution of the eye). But astronomers need even better resolution for quasars. This means building interferometers with longer baselines. Such baselines are now growing out from Jodrell Bank to lengths of 100 km. Remotely-controlled radio-telescopes are sited at the new outstations, joined by a radio link to make phase-stable interferometers. Long periods of observation with such systems allow astronomers to make detailed maps with a resolution reaching down to $\frac{1}{20}$ arc second. This is still not good enough for the innermost parts of quasars, and to study these inter-ferometers have become international, using baselines of thousands of

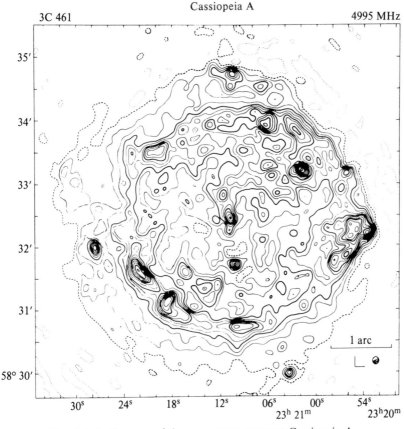

FIG. 8.1 Radio map of the supernova remnant Cassiopeia A.

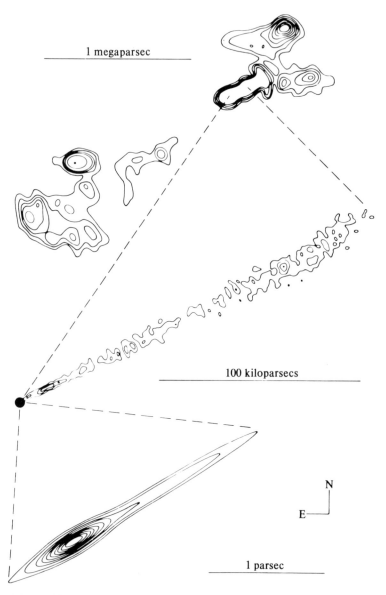

FIG. 8.2 Double-lobed radio source associated with the galaxy NGC 6251. Three stages of progressively higher spatial resolution are shown, with the top figure showing structure as large as several megaparsecs and the bottom showing detail on scales as small as parsecs.

kilometres. We now find a group of, say, five telescopes, in five different countries, all observing one source together, recording the results on magnetic tape, and sending their recordings to one centre for correlation and analysis. The final result is a map with resolution of about 1 milliarcsecond, and a scientific publication reporting the results with typically a dozen names on it. The whole picture requires all three stages—the aperture synthesis telescope for 1 arc second resolution maps, and the two other stages for the fine detail.

These stages of angular resolution are shown in Fig. 8.2. Here an extragalactic radio source is shown, with the typical double-lobed structure extending over several minutes of arc. Inside this there is a very well aligned jet structure, pointing towards the main radio lobes. Inside this, on the smallest scale, there is a quasar, with a jet structure only a few milliarcseconds across. The range of scales is astonishing, from megaparsecs down to less than one parsec. The physical circumstances which give rise to release of energy in such objects are now being unravelled: there appears to be an enormous rotating massive object at the centre of the quasar, attracting material which falls in a disc around the rotation axis. Explosive releases of energy, along the axis, can actually be seen changing the shape of the inner components of such sources in only a few weeks.

Radioastronomy therefore has the two essential ingredients—available information, and available techniques. Furthermore, it can operate from the ground. Fig. 8.3 shows the atmospheric transparency over the electromagnetic spectrum from radio to gamma-rays. For those wavelengths where the atmosphere absorbs significantly, the line shows the height where the intensity is halved. Above this height observations are feasible, below

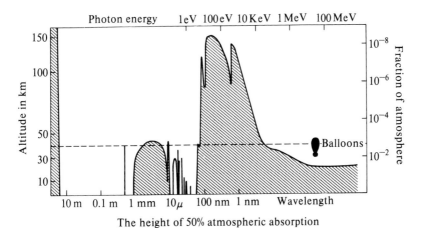

The height of 50% atmospheric absorption

FIG. 8.3 The transparency of the atmosphere to electromagnetic radiation from the radio region to gamma rays.

they are difficult or impossible. Radio extends from wavelengths of 15 m, where the ionosphere acts as a complete block, to below 1 cm, where oxygen and water vapour absorption bands start. Visible light is tucked inbetween the far infrared and ultraviolet, both of which are seriously absorbed. Our next example takes us to the extremely short wavelengths at the right of the diagram, where ground-level observations are impossible. This is the region of gamma-ray and X-ray astronomy.

Getting above the atmosphere requires balloons (one is shown in the diagram at a typical altitude of 40 km) or satellites, which are more expensive but which can work for years rather than days, the present limit of balloon flights. Modern balloons are very beautiful creations of polythene sheet, lifting payloads of half a tonne containing experiments, radio equipment for telemetry, and a parachute for recovery. Stabilized platforms which can direct telescopes within an accuracy of a few arc seconds are now available.

Gamma-rays have energies greater than $\frac{1}{2}$ MeV, where a spectral line is expected from electron–proton interaction. There are, at present, no spectral lines observed above this energy. The continuum spectrum is generated by cosmic rays, high-energy particles which permeate the whole of our Galaxy. The sky is relatively featureless in present gamma-ray surveys. The Sun generates no gamma-rays, and is not visible at all. Centaurus A, a galaxy already known as a radio and X-ray source, does appear, and so do four pulsars. Not, it seems, an outstandingly productive part of the spectrum. Nor is it technically easy. Gamma-rays are usually detected using spark chambers, apparatus which is familiar to nuclear physicists. The same cosmic-ray particles that make gamma-rays in space also make unwanted tracks in spark chambers, and they must be distinguished from the gamma-ray tracks and rejected. The apparatus needed, even for the present gamma-ray telescopes, is formidable, and there seems little chance of developments comparable to those of radioastronomy.

We have now looked briefly at the two ends of the spectrum; as we move in towards the centre we shall find richer and richer fields. Extending from the radio we have the millimetre waves, down to just below 1 millimetre wavelength (300 Gigahertz). In this region the broad-band, or continuum, radiation of radio sources such as quasars is too weak to be interesting, but there is instead a wealth of spectral lines. In the radio band there are important spectral lines of atomic hydrogen (at 21 cm), and hydroxyl (at 18 cm), but the concentration of lines increases sharply in the shorter radio and millimetric bands. There are lines from simple molecular species such as carbon monoxide and ammonia, from basic organic species such as methyl and ethyl alcohol and formic acid, and there are lines from complex and less familiar organic molecules such as cyanobutadyne and cyanohexatryne. The basic ingredients of living matter are present in condensing gas clouds, such as the Orion Nebula (Plate 1) where star formation is actually occurring. The full explanation of this cosmic chemistry now awaits the development of appropriate observing techniques (which incidentally are also needed by the telecommunications industry). This is a region where radio shades into optics, and new techniques are needed to develop waveguides, detectors, and antennas. There is a proposal for an SRC millimetre wave telescope which will explore this new band down to the limit of ground-based techniques. Fig. 8.4 shows in more detail the atmospheric attenuation at mountain-top heights. The last reasonable gap, at 0.8 mm (400 Gigahertz),

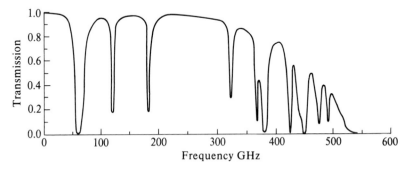

FIG. 8.4 Atmospheric transparency in the millimetre wavelength range at 4000 m altitude for 1 mm of water vapour.

is the target for the new telescope. It will be 15 metres across (Fig. 8.5); it must maintain a surface accurate to about 50 μm, and so must be placed in a windproof and thermally controlled shelter looking remarkably like the dome of an optical telescope.

Between millimetre waves and the infrared lies a region which cannot be explored from the ground, or even from balloons, and for which good techniques do not yet exist. There will perhaps one day be a far infrared satellite exploring molecular spectral lines in this part of the spectrum, but as yet we do not know how to make it.

Moving in from the short wavelengths, we come to a very exciting region—the X-ray region. In 1962 Giacconi, following the pioneering work of Friedman, discovered the cosmic X-ray source known as Scorpius X-1. This was a thousand times stronger than the Sun, and showed that it was worthwhile to develop X-ray observing techniques from rocket-borne to satellite-borne detectors. The first X-ray telescope containing focusing surfaces was constructed by the Mullard Space Science Laboratory and installed in the satellite Copernicus. Later came the satellites *Ariel 5*, and *HEAO-B*.

X-ray sources are mostly thermal: they are very hot, about a million degrees absolute, and often quite small. Because of the high temperature the source radiates peak emission in the X-ray band, and negligible emission in the visible. Such conditions are generated in a certain type of close binary star where matter is being transferred from one partner to the other. X-ray emission is produced when the latter is a condensed neutron star, whose gravitational field sucks gas from the outer layers of a normal stellar partner. The gas falls in onto the tiny neutron star, perhaps only a few kilometres across. Apart from the X-rays, there would be very little indication of the existence of the neutron star component of these binary systems.

There are at least 300 X-ray sources known, about 100 of them extra-galactic (see Fig. 5.5). I will pick on one, a galactic source discovered at Christmas 1974, in the constellation of Centaurus, and therefore popularly known as Cen X-mas. This is a variable transient source similar to TrA X-1

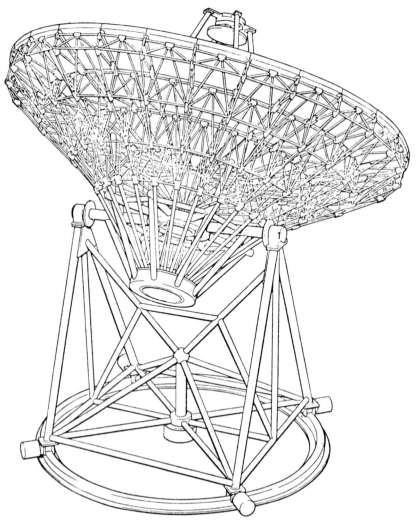

FIG. 8.5 Sketch of the proposed UK millimetre wave telescope.

described by Professor Pounds (p. 104). An optical counterpart was dis-covered in a search for a similarly varying visible object. Furthermore, Cen X-mas was found to be the first slow X-ray pulsar, giving regular variations with a period of $6\frac{3}{4}$ minutes. Adding the optical to the X-ray results we obtain a fairly complete picture of this binary system, including a measurement of its distance.

The X-ray wavelengths contain very few spectral lines. One that is observed comes from very highly ionized iron and is found in the gas filling clusters of galaxies. Another is believed to be a cyclotron resonance of electrons in the strong magnetic field at the pole of a magnetized neutron star. But, generally, X-ray spectra are continuous and relatively featureless.

We have started from the two ends of the spectrum, and are closing in on the rich central regions. Moving from X-rays to the ultraviolet, we cross a region where a hydrogen spectral line, the Lyman alpha line, is dominant. Shortwards of this line the interstellar gas absorbs radiation strongly. Despite this there is doubtless much of interest to be learned between 100 and 1 nanometres wavelength. This is a relatively unexplored region. Longwards of 100 nanometres lies the very rich ultraviolet region, full of spectral lines and full of energy: the peak radiation of hot stars falls in this band. Furthermore, we have the techniques for exploring it, even though we must observe from above the atmosphere. This is the region covered by the satellite *IUE*, the *International Ultraviolet Explorer* and other, earlier satellites.

FIG. 8.6 An artist's impression of the International Ultraviolet Explorer satellite with an ultraviolet telescope on board. (Courtesy of *Nature*.)

IUE has the following main characteristics. It covers 115 to 320 nanometres (in two bands), using spectrographs with resolution 0.02 and 0.6 nanometres. It is in a geosynchronous, elliptical orbit, with radius between 46 000 km and 26 000 km. It operates continuously, 16 hours per day being available from USA and 8 hours from Europe. It weighs 700 kilograms and requires 210 watts of power. It uses a television camera as a detector, with a converter to turn ultraviolet into visible light.

How does an astronomer use *IUE*? He travels out to an observatory near Madrid, and looks at a television monitor as though he were using any modern optical telescope. He can steer the satellite on to the required target star by remote control, set up the spectrograph, and after the required exposure time record a spectrum as though it was being taken on a photographic film. The high-resolution spectra are folded, using an echelle grating, and appear packed into the screen as in Fig. 8.7. Here we have the spectrum of a fairly cool star, Capella, which has bright emission lines (hydrogen at 1226 Å, silicon at 1206 Å, etc.). These come from a hot chromosphere outside the cool surface of the star. Such a spectrum is transmitted from the satellite as a television scan, and it can be analyzed immediately through a data-processing computer to produce a fully corrected and linear spectrum, as in the example of the quasar 3C 273 (Fig. 8.8). This spectrum has a red-shift of 0.158. Its main value is in calibrating the ultraviolet spectra of quasars with much larger red-shifts, so that lines such as Lyman alpha appear in the optical range.

The ultraviolet region requires satellites to get high enough above the

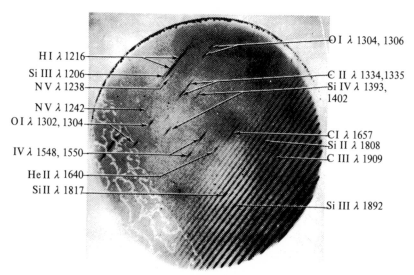

FIG. 8.7 High-resolution spectra from the International Ultraviolet Explorer satellite as they appear on the observing screen at the satellite tracking station. (Courtesy of *Nature*.)

Earth's atmosphere. Infrared, at the opposite end of the optical range, can be observed from the Earth's surface. But even here there is a need to keep above most of the water vapour, and it is a great advantage to build infrared telescopes on high mountain tops. A new British infrared telescope, the UKIRT, is just starting observations on Mauna Kea at 14 000 feet elevation on the island of Hawaii. This telescope is very like a conventional optical telescope. With a mirror 3.8 metres in diameter it is among the largest telescopes in the world, and only differs from optical telescopes in its light-weight construction which is permitted by the relaxed tolerances allowed by the longer infrared wavelengths.

We end where we began, in the optical band. Despite all the new information available from the 'high-energy' astronomies observing at shorter wavelengths, the fact remains that the optical band is far and away the richest in available information. Furthermore, the available instrumental techniques are still improving rapidly. Photographic plates, themselves greatly improved recently in response to the demands of astronomers, are now being superseded by television detectors with greater quantum efficiency and linearity. There has been a transformation in telescope design through the application of modern control engineering. Above all, there has been a liberation of attitude towards the location of observatories, so that it is no longer necessary to build good telescopes like the Isaac Newton in bad climates such as that of the UK. The enormous opportunities for optical astronomy have led to the proposal by the Science Research Council of a new optical observatory for Northern Hemisphere observations. I shall

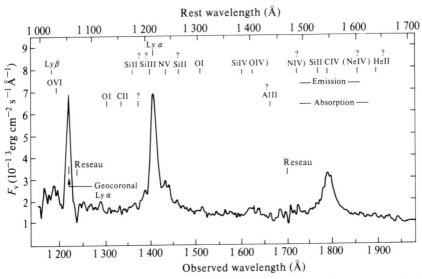

FIG. 8.8 Fully corrected ultraviolet spectrum of the quasar 3C 273 from the IUE satellite. (Courtesy of *Nature*.)

spend some time outlining this proposal since it should be complete within a few years.

The essential requirement is to be able to observe the faintest, most distant visible objects in the Universe. These are, all too often, the objects for which the X-ray- and radioastronomers most urgently need more information, and particularly information on their spectra. This requires a very large telescope on a very good site. Not only must the site be as free from cloud as possible, but it must also have a steady atmosphere which does not blur star images into large discs. The efficiency of a spectrograph and telescope combination is degraded in proportion to the size of the seeing disc: commonly the disc is 2 or 3 arc seconds across, while on an excellent site it is less than 1 arc second for at least half the observing time. For satellites above the atmosphere, such as the proposed Space Telescope, there is no such limitation: indeed, if space telescopes could also be made very large and reasonably economical most optical work would be carried out by satellite. But that is not the case, and we have therefore looked very

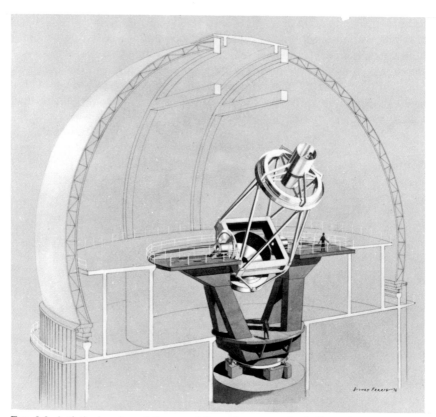

FIG. 8.9 Artist's drawing of the new 4.2 m telescope to be built at the new Spanish international observatory on the island of La Palma in the Canary Islands.

carefully for the best site on the ground. The island of La Palma, in the Canary Islands, has been chosen after a thorough survey, and it has been made available by Spain for international use by astronomers. The observatory site will shortly become the Roque de los Muchachos Observatory, and it will have telescopes belonging to UK, Sweden, Denmark, and Germany.

The new observatory is at 2500 metres altitude, which is above the inversion cloud layer common in the Canary Islands. Continuous monitoring over a complete year has established that it qualifies as the best site in Europe, both in respect of freedom from cloud and in quality of star images, and it is possibly the best site in the whole of the northern hemisphere. The arrangements for international participation are a model of co-operation: the Spanish side will provide the site, the road, and all essential services in return for the training of Spanish astronomers and the use of 20 per cent of the time of all telescopes. The British telescopes will be a new 1-metre, the 2.5-metre Isaac Newton, and a new 4.2-metre telescope. The latter will be the principal telescope for co-operative work with the X-ray- and radioastronomers. The 1-metre telescope is already under construction, the Isaac Newton at Herstmonceux, Sussex has been closed for reconstruction work before moving to La Palma, and the 4.2-metre is in the process of design. A glass–ceramic blank is already available for the primary mirror. This new telescope will use an alt-azimuth rather than an equatorial mount.

Optical astronomy has enjoyed an outstandingly successful period recently. The Anglo-Australian telescope has led the way in the use of new techniques of automatic telescope control and new detection systems. To pick on one single result: the discovery of the optical pulses from the Vela pulsar, and the subsequent identification with the faintest object yet seen, were only made possible by these new techniques. New television detection systems are approaching the ultimate in sensitivity, where every photon is counted.

Also in Australia is a new British telescope of first importance: the 1.2-metre Schmidt survey telescope. This is taking survey photographs of the whole of the southern sky, and achieving one or two magnitudes fainter than the well-known Palomar survey of the northern sky. With the addition of an objective prism, which spreads out each star image into a tiny spectrum, objects with a distinctive spectrum can easily and automatically be picked out. In particular, the quasars on each plate can be distinguished from ordinary stars and galaxies. It now seems that each plate 6 degrees square will allow us to discover as many new quasars as were known by all the methods used before this telescope started operation.

You will recognize my enthusiasm for the programme of ground-based optical astronomy to which all British astronomers are dedicated at this time, and you may be wondering whether ground-based optical astronomy is becoming out-dated by the Space Telescope. As you know, this is a large telescope, 2.4 metres diameter, planned for launch in 1983. Operating above the atmosphere, it has two advantages—it will work in the ultraviolet, and it will not suffer from 'bad seeing'. Star images will be only about $\frac{1}{6}$ arc second across, and the detection sensitivity will be correspondingly improved. This telescope will undoubtedly work on fainter objects than any other telescope. British astronomers aim to take full advantage of any opportunity that may present itself of using the Space Telescope, either by direct application to agencies in the USA or through the 15 per cent share to which the European Space Agency is entitled.

The only trouble with the Space Telescope is that there will be only one. European astronomers will have 15 per cent of the time, and the telescope will operate 24 hours a day, but the time will be so precious that it will be reserved only for those observations which cannot be done from the ground. However, time can often be traded for sensitivity: a whole night on the 4.2-metre will, for most purposes, be much more useful than a single hour on the Space Telescope. The Space Telescope is a very special instrument, but its work must be supplemented by the more versatile and more readily-available ground-based telescopes.

In this survey I have worked through the electromagnetic spectrum, and now look briefly at other astronomical studies which lie outside it. They are limited, but important. First, for our nearest neighbours in space, the Moon and the planets, man, or his apparatus can go there. We are forced to admit that this applies only to the solar system, but nevertheless there is no substitute for genuine moon-rock samples, or for direct photographs of the planetary surfaces described by Dr Hunt.

Various kinds of particles reach us from outer space. Cosmic rays contain information on interesting processes of particle acceleration, but nothing about their origin since they arrive isotropically. Neutrinos flow through our laboratories, and are probably full of information—but unfortunately they are very difficult to detect. There is a classical experiment under way which has been described by Professor Blackwell which aims to detect about one neutrino a day from the Sun. But there is not much chance at present of detecting them at a significantly greater rate. Material particles are not, generally speaking, very informative about the Universe.

The last possible extension is to detect gravitational waves. I include this possibility because this year is the Einstein centenary year, and because this year has also seen the first evidence that gravitational waves exist. The change in orbital period recently found in a unique binary-star system containing a pulsar can only be accounted for by a loss of energy through gravitational waves. But it is another matter to detect such waves directly on Earth, and all attempts so far have failed. It is a further case of plenty of information, but no technique available even to detect the signals, let alone discover where they come from.

In this chapter I have outlined some of the tools available to the modern astronomer. We are indeed fortunate to live in an age which has seen Einstein, the exploration of the electromagnetic spectrum, and man's first steps on the Moon. Astronomy contains within itself almost the whole range of physics; further, it catches the popular imagination as does no other branch of science. It is the most inspiring of the sciences, and the most useful in education. The next few decades will find British universities singularly well equipped in all observational methods—and especially in the optical field which remains the richest in opportunities.

Index

Absolute event horizon, 132
Accretion disc, 100, 126, 127
Acetylene, 167
Active galaxies, 24, 112
Amalthea, 172, 173
Ammonia, 167
Andromeda galaxy, 5, 6, 17, 32, 35
Anglo-Australian telescope, 70, 80, 104, 193
Angular diameter, 78–80, 83
Angular resolution, 182
Antimatter, 59
Aperture synthesis, 183
Area increase principle, 139
Ariel 5, 95, 97, 102, 104–5, 112–13, 118, 187
Atmospheric turbulence, 78

Balloons, 102, 186
Baryon number, 57
Beckenstein, J. D., 140
Big bang, 4, 9, 12, 14, 32, 42, 111, 138, 141
    cold, 59
    hot, 56–7, 59, 60
Big crunch, 10, 34, 58
Binary pulsar, 128
Black-body radiation field, 10
Black hole, 27, 29–32, 38, 56, 102, 121, 132, 139
Bondi, H., 9
Burbidge, G., 25, 49

3C236, 27, 29
3C273, 113, 117, 190
Callisto, 173, 175, 178
Carbon-burning, 51
Carbon dioxide, 149, 153–6, 158, 163
Carbon monoxide, 149, 153, 156, 167
Cassiopeia A, 183
Centaurus A, 29, 31, 113, 186
Centaurus X-3, 98, 100, 102
Chandrasekhar, S., 122
Charge number, 57
Classical novae, 105
Clusters of galaxies, 3, 6, 32, 48, 65, 109, 189
Cold big bang, 59
Conserved quantum numbers, 57
Coma cluster, 7, 32, 109

Cosmic censorship, 138
Cosmic rays, 52, 194
Crab nebula, 94–5, 98, 100, 108
Craters, 148, 164, 175
Critical density, 10, 37
Cygnus A, 24
Cygnus loop, 55, 109–10
Cygnus X-1, 100, 102, 125–6

Dark companions, 90
Degenerate matter, 52, 122
Deuterium, 13, 56, 58
Deimos, 163
Disc stars, 47
Doppler shift, 7, 28
Double radio sources, 24, 27–8, 185
Dust
    interstellar, 46, 70
    planetary, 150, 161
    storms on Mars, 160–1
Dwarf novae, 103

Effective temperature, 75, 77, 85
Einstein observatory, 111, 116, 118, 119
Electron, 123
Electron lepton number, 57
Element abundances, 40, 42
Elliptical galaxies, 18, 45
Ergosphere, 139
Escape velocity, 121
Ethane, 167
Europa, 173, 175–6

Galactic nucleosynthesis, 60
Galaxies, 16
    active, 24, 112, 189
    clusters of, 3, 6, 32, 48, 65, 109, 189
    elliptical, 18, 45
    formation of, 32
    irregular, 22, 45
    properties of, 6, 16
    radio, 24
    Seyfert, 113, 115
    spiral, 17, 45
Galileo, 172
Gamma-rays, 93, 186

Gamow, G., 12, 42
Ganymede, 173, 175, 177
General relativity, 10, 37, 129, 134
Globular clusters, 41, 43, 46, 61–2, 100
Gold, T., 9
Gravitational waves, 194
Great red spot, 169–70

Halo stars, 47
Hawking, S., 14, 137, 140
Heavy elements, 41, 44, 46, 50, 53, 57, 61–2
Helium, 12, 41, 44, 46, 50, 56–9, 68, 122, 165, 167
Helium-burning, 51
Hercules X-1, 100, 102
Hipparcos, 89
Horizon (*see* Absolute event horizon)
Hot big bang, 56, 57, 59, 60
Hoyle, F., 9, 49
Hubble, E., 6, 7
Hubble's law, 7, 8, 9, 34
Hydrogen, 41, 44, 50, 56–8, 68, 122, 165, 167
Hydrogen-burning, 50

Ice on Ganymede and Callisto, 175
Inert gases, 149
Infrared, 83, 86, 191
Infrared flux method, 83
Intensity interferometer, 81
International Ultraviolet Explorer, 87, 189
Interstellar dust, 46, 70
Interstellar medium, 52, 64, 72, 109
Intracluster gas, 48, 111
Io, 166, 173–4
Iron, 48–9, 65, 68, 111–12, 189
Irregular galaxies, 22, 45
Isotopic abundances, 48

Laplace, P. S., 121
Lepton number, 57
Light cone, 129, 130–1
Lightning, 150
Local Group, 6, 32

Jeans, Sir James, 38
Jupiter, 41, 146, 165, 166
   atmosphere, 167, 170
   magnetosphere, 167
   ring, 171

M82, 24, 113, 125
M87, 24–5, 31–2, 136
Magnetic field, 102, 148, 165–6, 175, 189
Main sequence, 20
Mars, 146, 155, 166, 178
   atmosphere, 155–6, 158, 163
   water-sculpted forms, 161
Mariner, 146–7
Mercury, 146, 166
Meteorites, 41, 48

Methane, 167
Microwave background radiation (see black-body radiation field), 57
Millimetre wave telescope, 186
Missing matter, 10, 13, 38
Muon lepton number, 57

NGC 4151, 113
NGC 6251, 28, 184
Neptune, 146, 177
Neutrinos, 38, 53, 59, 60, 75, 194
Neutron, 40, 49, 52, 58, 123
Neutron star, 55–6, 102–3, 123–4, 188–9
Nitrogen, 165
Nuclear reactions, 49, 53, 73, 75
   carbon-burning, 51
   helium-burning, 51
   hydrogen-burning, 50
   oxygen-burning, 51
   silicon-burning, 51
Null curves, 132

Occultation, 80
Orion nebula, 21, 73
Oxygen-burning, 51

Penrose, R., 14, 121
Penzias, A., 11
Phobos, 163
Pioneer, 146, 149, 153, 165, 167
Phosphine, 167
Planetary spacecraft, 146
   Mariner, 146–7
   Pioneer, 146, 149, 153, 165, 167
   Venera, 146, 149
   Viking, 155, 160, 163, 177
Principle of equivalence, 134
Properties of galaxies, 6, 16
Proton, 40, 49, 58, 123
Protostars, 72–3
Pulsars, 55–6, 108–9, 124, 181, 194

Quantum theory of gravity, 14
Quasar, 27, 29, 35, 113, 185, 193

Radio galaxies, 24
Radioastronomy, 182
Red giant, 122, 124
Red-shift, 7, 24, 34, 190
Relativity, 128
   general, 10, 37, 129, 134
   special, 129, 134

Satellites, 186
   Ariel 5, 95, 97, 102, 104–5, 112–13, 115, 187
   International Ultraviolet Explorer, 87, 189
   Space Telescope, 38, 70, 79, 178, 192, 193, 194
   Uhuru, 94–5, 98, 104, 109, 112
Saturn, 41, 146, 171

Schmidt survey telescope, 193
Scorpius X-1, 94, 100, 186
Seyfert galaxy, 113, 115
Shipley, H., 6, 16
Silicon-burning, 51
Singularity theorems, 14, 136
Sodium, 173
Solar oscillations, 81, 92
Solar-neutrino experiment, 53, 75
Space Telescope, 38, 70, 79, 178, 192, 193, 194
Spanish International Observatory, 192
Special relativity, 129, 134
Speckle interferometer, 79
Spectrum analysis, 68
Spiral arms, 22
Spiral galaxies, 17, 45
Star-formation, 22, 61, 63
Stars, 40, 68
   binary pulsar, 128
   disc, 47
   evolution of, 76
   formation of, 22, 61, 63
   halo, 47
   neutron star, 55–6, 102–3, 123–4, 188–9
   novae, 103, 105
   protostar, 72–3
   pulsar, 55–6, 108–9, 124, 181, 194
   red giant, 122, 124
   supernovae, 21, 54–5, 123
   white dwarf, 103, 122, 124
   X-ray binary, 98, 101, 188
Steady-state cosmology, 12
Stellar evolution, 76
Stellar interferometer, 80

Stellar parallax, 88
Strong nuclear interaction, 57
Sulphuric-acid, 153
Supernovae, 21, 54–5, 123
Supernova remnant, 55, 108–9, 183
Synchrotron radiation, 24, 108, 165

Tidal disruption, 31
Tidal distortion, 135
Tides, 173, 175
Timelike curves, 131
Titan, 177

Uhuru, 94–5, 98, 104, 109, 112
Ultraviolet, 86–7, 93, 189
Uranus, 146, 171, 177

Venera, 146, 149
Venus, 146, 148, 166, 177
   atmosphere, 151
Viking, 155, 160, 163, 177
Volcanoes, 150, 156, 162, 173–4
Voyager, 146, 165, 167–8, 172, 175

Water-sculpted features on Mars, 161
Water-vapour, 149, 156, 167
Weak nuclear interaction, 57
White dwarf, 103, 122, 124
Wilson, R., 11

X-ray binary stars, 98, 101, 188
X-rays, 93, 187
X-ray transients, 103

Zodiacal light, 84